LIMITING NOISE FROM PUMPS, FANS AND COMPRESSORS

LIMITING NOISE FROM PUMPS, FANS AND COMPRESSORS

I Mech E CONFERENCE PUBLICATIONS 1977-10

Conference sponsored by the
Fluid Machinery Group of
The Institution of Mechanical Engineers

London, 25 October 1977

Published by
Mechanical Engineering Publications Limited for
The Institution of Mechanical Engineers
LONDON

First published 1977

This volume is complete in itself.

There is no supplementary discussion volume.

ISBN 0 85298 380 8

Printed and bound by The Burlington Press, Foxton, Royston, Hertfordshire

CONTENTS

Occupational deafness, by S.J. Karmy   1

The need for the classification of pump noise and vibrations,
by D. France   9

Development of a standard, silenced, oil injected screw type, air compressor
plant, by G.L. Collier   15

A user's view on the control of noise from pumps, fans and compressors,
by J.B. Erskine   23

Recent work by CONCAWE on the measurement of noise emission from
large sources, by K.J. Marsh   39

Current industrial pump and fan fluid-borne noise level prediction,
by W.M. Deeprose and A.N. Bolton   43

Noise control strategy for manufacturers and users, by M.F. Russell   51

Fan noise research and its implication for noise level predictions,
by B.D. Mugridge   59

A pump manufacturers approach to noise emission limitations,
by J.K. Macdonald and B. Reil   65

OCCUPATIONAL DEAFNESS

S.J. KARMY, BA(Electron.), MSc, MIOA
Institute of Sound and Vibration Research, University of Southampton,
Southampton.

The MS of this paper was received at the Institution on 2nd June 1977 and accepted for publication on 19th July 1977

SYNOPSIS   The problem of industrial deafness has been well known for over one hundred years, but it
is only within the last decade or so that it has been accurately related to industrial noise in
terms of equivalent continuous noise level.   This paper describes the ear and noise-induced hearing
loss, the social consequences of such deafness, together with the current standard measure of
industrial noise taken to assess hazard to hearing. A general outline of an overall approach to
hearing conservation is also presented in terms of the Department of Employment's Code of Practice
on noise exposure and the current legal situation.

INTRODUCTION

1.   It would appear that the first author to
describe noise induced hearing loss was Pliny the
Elder (ref. 1), who produced his work "A Natural
History" in the first century A.D.   Therein is
recorded the observation that people living near
to the large waterfalls of the Nile eventually
became hard of hearing.

2.   Francis Bacon (ref. 2) refers to Pliny in
his treatise "Sylva Sylvarum" produced in 1627,
and makes comments of his own concerning the
adverse effects excessive noise can have upon
the hearing. Similarly Ramazini (ref. 3) in 1713
describes noise deafness among coppersmiths, and
in the same century Admiral Lord Rodney is reported to have been deafened for a fortnight by
cannon fire in 1782.

3.   However almost another fifty years passed
before a more complete account of noise induced
deafness appeared in the literature, this being
the work of Fosbroke (ref. 4) in 1831, who was
concerned about hearing levels in the blacksmith
trade. Experimental evidence was published in
1886 by Thomas Barr (ref. 5), who measured a
differential in hearing levels between employees
of several types of noisy trade. In particular
he noted that 75% of boilermakers whom he had
tested showed signs of deafness.

4.   Despite such a historical pedigree, it is
only within the last decade that industry has
made a concerted effort to prevent the occurrence
of noise induced hearing loss in its workforce.
It is now possible to describe many of the mechanisms by which excessive noise can damage the
hearing, but despite a large research effort,
many areas of uncertainty remain. However, the
adverse effect that excessive noise has upon the
hearing of exposed persons is well documented,
and the onus of reducing this hazard is placed
squarely upon the shoulders of the employer.

THE EAR

5.   In order to discuss noise induced hearing
loss further, it is necessary to describe,
briefly, the passage of sound through the ear.

6.   Figure 1 shows a longitudinal section through
the human ear. The three major divisions are
clearly marked, these being the outer ear, the
middle ear and the inner ear.

7.   The incoming sound wave is collected by the
pinna, and passes down the ear canal to impinge
upon the eardrum. The sound wave is then transmitted across the middle ear space via the
ossicles, a linked system of three tiny bones, to
the oval window of the inner ear. The tympanic
membrane (eardrum) and ossicles together function
as an impedance matching device to allow the
energy in the airborne sound wave to be transmitted efficiently through the coils of the
liquid-filled cochlea of the inner ear. Excess
sound energy eventually dissipates itself back
into the middle ear cavity through the membranous
covering of the lower opening in the cochlea; the
round window.

8.   Habitual exposure to hazardous levels of
noise has the most damaging effect upon the inner
ear.   That is not to say that excessive noise cannot damage the middle ear. Large, abrupt pressure
changes brought about, for example, by an explo-
sion or sudden atmospheric decompression, can
cause rupture of the tympanic membrane or disloca-
tion of the ossicles. A wide range of surgical
techniques exist, however, to repair the results
of such trauma, if the damage is not sufficiently
minor to allow natural healing to occur (Colman,
ref. 6).

9.   At the present time, damage caused to the
inner ear is not repairable, and can be caused by
prolonged exposure to hazardous levels of industrial noise.

10. To discuss the locus of such a noise induced hearing loss, the structure of the cochlea must be examined in greater detail. Figure 2a shows a three dimensional diagram of the cochlea. As can be seen, the coiled tube is partitioned along its length, dividing the cochlea into two liquid-filled volumes; the scala vestibuli, in contact with the oval window, and the scala tympani, in contact with the round window. A small gap, the helicotrema, resulting from the cochlear tube being slightly longer than the partition, allows the two scala to communicate. Figure 2b shows the partition in greater detail and reveals that it is itself a duct. Pressure waves set up in the cochlear fluids by the vibration of the final bone in the ossicular chain, the stapes, cause the floor of this duct, the basilar membrane, to flex. This in turn causes movement of the hair cells, resulting in deflection of the sensory hairs which protrude from the cuticular plate on the top of each hair cell. Deflection of the sensory hairs causes a potential to be generated by the hair cell, a potential which is detected by an underlying cochlear neurone. The resulting electrical signal is then transmitted to the higher auditory centres via the auditory nerve, and gives the sensation of hearing.

NOISE DAMAGE TO THE INNER EAR

11. It has been demonstrated that the hair cells residing in the Organ of Corti become fatigued by constant exposure to excessive noise, and are destroyed (Hunter Duvar et al, ref. 7). Thus the mechanical movement of the basilar membrane which arises in response to audible pressure waves can no longer be coded into neural impulses in the auditory tract, and deafness exists.

12. As can be seen from figure 2b, two distinct populations of hair cells exist, inner hair cells, and outer hair cells. As well as being spatially separated, the two populations differ slightly in the type and arrangement of sensory hairs.

13. Noise damage appears to occur to both the inner and outer rows of hair cells, although the outer rows are the first to be damaged by trauma of this type (Bohne, ref. 8; Dieroff et al, ref. 9). In the case of row depletion of the hair cells due to overstimulation by noise, there exists a reasonable correlation between the place of loss on the basilar membrane and the section of basilar membrane thought to be responsible for transducing the damaging acoustic signal (Hawkins et al, ref. 10). Bohne (ref. 8), however, does quote evidence showing that damage can extend beyond the section of basilar membrane responsible for transducing the damaging signal.

14. It is thought that loss of hair cell function can still occur even if the hair cell remains present on the basilar membrane (Hunter Duvar et al, ref. 11). This loss of function is generally attributed to damage of the ultrastructure of the hair cell. Spoendlin et al (ref. 12) reports the permanent bending of the sensory hairs of the hair cells as an example of damage caused by excessive noise exposure.

15. Such functional loss without actual hair cell degeneration causes practical difficulties for research workers who attempt to define the role of the inner and outer populations of hair cells.

16. Ward et al (ref. 13) and Eldredge et al (ref. 14) have reported that outer hair cell damage can occur without a permanent threshold shift (hearing loss) being sustained. This would indicate that the inner hair cells alone mediate the normal threshold of hearing. Unfortunately these findings have proved difficult to replicate (Hunter-Duvar et al, ref. 7), Ward (ref. 15).

17. The reason for the existence of two populations of hair cells might be explained by a suggestion made by Kiang et al (ref. 16). They postulated that the electrical responses from the inner and outer hair cells interact in some way so as to produce the otherwise unexplicably fine response characteristics exhibited by the auditory nerve fibres.

18. This point is relevant in a discussion of noise induced hearing loss, for it could explain the lack of auditory discriminatory ability experienced by sufferers of this disability. This phenomenon is typified by the "Cocktail Party" effect, in which a person with a noise induced (sensori-neural) hearing loss experiences difficulty in picking out a certain conversation from a background of noise or other conversations. In all probability it is for this reason that a hearing aid will give only slight relief to a person suffering from a noise induced loss. A hearing aid can at best only replace the lost gain in a damaged hearing system, and not the other, more complex functions of the inner ear.

MECHANISMS OF NOISE DAMAGE

19. It is certain that not all noise induced damage can be explained by simple mechanical over-stressing. Spoendlin (ref. 12) concluded that there existed a critical level of approximately 130 dBSPL above which damage in the cochlea was mainly mechanical in nature, a damaging exposure time being measured in minutes. Below this level the time taken to damage the hair cells was much longer, taking place as the result of a metabolic alteration resulting in the eventual destruction of the hair cells.

20. If exposed to moderate levels of industrial noise, for example 93 dBA throughout each working day, the time taken to develop a sizeable hearing loss would be measured in years for the majority of the human population. It must be remembered that not all individuals are equally susceptible to noise induced hearing damage. Work has been proceeding for many years aimed at producing a screening test to find those individuals who have hearing susceptible to noise damage, but without success. The problem still has to be viewed on a statistical basis.

21. The phrase "metabolic alteration" mentioned earlier is really a blanket term to cover scientific uncertainty. One possibility, however, that of oxygen lack, has been ruled out. Despite studies by Jansen (ref. 17) who produced evidence of noise induced vasoconstriction of small blood vessels, blood flow to the inner ear has been shown to remain unaffected until levels of 120 dB SPL are produced, when the blood flow into the ear actually increases. (Perlman et al, ref. 18). Thus Ward (ref. 19) concludes "the (previously observed) decrease in oxygen tension associated with moderate exposure must indicate only an

increase in oxygen consumption, rather than a decrease in oxygen supply". Other hypothetical mechanisms for noise induced interference with metabolic processes include disruption of enzyme activity within the cochlea, and the levels of glycogen within the hair cells (Engström et al, ref. 20).

POST EXPOSURE NOISE DAMAGE

22. It is of interest to consider whether or not disintegration of cochlear structures continues after noise exposure has ceased. It would appear that this depends upon the amount of damage initially sustained by the hair cell during exposure. Serious damage will result in the continued disintegration of affected hair cells for several months, but animal studies have shown that slight hair cell damage will recover over a period of a few months. This agrees with the observation that often the hearing of a noise exposed employee may well recover slightly over a period of six months after leaving the hazardous noise environment.

23. Loss of hair cells is usually followed by loss of a percentage of the cochlear neurones responsible for the innervation of that particular area of the basilar membrane. Spoendlin (ref. 12) summarises the topic by stating that degeneration of the cochlear neurones will only take place if both the inner and the outer rows of hair cells have disappeared. Even if the basilar membrane is stripped over a certain area, 10% of the population of the cochlear neurones will still survive.

24. The continued existence of some of the nerve fibres from the damaged areas of the basilar membrane is vital if the latest potential aid for the deaf, the cochlear implant, is to have any success. These devices, still in the highly experimental stages, code acoustic signals electrically, and then seek to stimulate the relevant cochlear neurones directly.

SOCIAL HANDICAP

25. The audiogram shown in figure 3 reveals the progress of a noise induced hearing loss sustained by an employee over a number of years. Although a developing noise induced deafness usually shows a recognisable pattern of loss of acuity with frequency, it is impossible to diagnose the presence of a noise induced hearing loss solely from the characteristics of the audiogram. Additional factors must be taken into consideration. Tests must be carried out to ensure that the site of the loss resides in the inner ear and not the middle ear. The medical history must be taken into account to dismiss the possibility that other agents such as ototoxic drugs, noisy hobbies, or a previously contracted illness were responsible for producing the inner ear damage. Even so, it is usually only possible to decide that a loss is noise induced "on the balance of probabilities".

26. As shown in figure 3, excessive noise affects the ear most severely at a frequency of 4 kHz. The reason for this has not been satisfactorily explained, although it is suggested that it is a result of the physical dimensions of the cochlea, and does not represent a particular weakness of the hair cells in the 4 kHz region. The notch first seen in the audiograms of figure 3 gradually deepens and widens towards the low frequencies with time. As can be seen from figure 3, the increasing hearing loss encroaches upon the frequency band used for speech. The first components of speech reception to be affected are the consonants, which unfortunately carry the majority of the information necessary for good speech intelligibility. As the hearing loss affects the 2.5 kHz to 3 kHz region, the listener can also experience some problems in identifying the speaker. The progress of a noise induced deafness is insiduous, developing in the majority of the population over a period of years. Often it is the family of the affected person who first become aware of the hearing problem, as a person suffering from a noise induced hearing loss will often start, unconsciously, to lipread in order to compensate for his deafness. This is particularly easy in the work environment where many conversations contain a routine element, and may be conducted in noise, with the associated raising of voice levels, and consequent assistance to the deafened employee. Eventually, however, such an employee may become a hazard to himself and others around him in the factory environment. Certainly he will be subjected to a feeling of increasing isolation, especially in his social activities which will probably reduce in number.

27. This increasing social handicap was perhaps first remarked upon by Barr (ref. 5) who entreated clergymen to ensure that any boilermakers in their congregation sat at the front of the church. Social handicap is a complex phenomenon, and attempts to measure and relate it to the observed hearing loss have been made without a great deal of success. Obviously quantification of social handicap would be invaluable in deciding upon the size of any compensation award to be made to a noise deafned claimant. Coles (ref. 21) points out that many social factors influence handicap, and hence "a close correlation with the degree of impairment is not to be expected". A diagramatic representation of those factors influencing social handicap is shown in figure 4. In studying the social handicap experienced by noise deafened weavers, Kell et al (ref. 22) demonstrated that the main problems were to be found in the various forms of social communication, and showed that the hearing loss averaged over 1, 2 and 3 kHz proved to exhibit the best correlation with the various aspects of social handicap that were measured, although the correlation was still not good.

28. A frequent consequence of noise induced hearing loss is that of tinnitus. Damage to the organ of corti can give rise to the generation of subjective noises. The sufferer hears whistling/ hissing sounds, etc., which in some cases can be so severe as to induce suicidal depression. Eradication of this tinnitus is not easy, even if total destruction of the organ of hearing is contemplated. Not every noise deafened employee, however, develops this symptom (Johnsson et al (ref. 23)).

PREVENTION OF NOISE INDUCED HEARING LOSS

29. The only remedy for a noise induced hearing loss is to prevent it from occurring in the first place. At the present time, industry is guided by a document issued by the Department of Employment, "A code of practice for reducing the exposure of employed persons to noise". More positive requirements in the form of legislation are expected soon, following the passing of the Health and Safety at Work Act, 1974 (Health and Safety Committee 1974). The Code of Practice sets

3

out those steps which are necessary to ensure that noise induced hearing loss does not become prevalent in a work force. It relies heavily upon the "equal energy concept" in setting the risk criteria. This principle states that the amount of permanent noise induced deafness is proportional to the total A-weighted energy received by the ear. Thus there exists a trade-off between the length of exposure and the intensity of the exposure, within certain limits. Use of the equal energy concept was first initiated by the United States Air Force (ref. 26) in 1956, but full experimental demonstration of the principle had to await the work of Robinson (ref. 27) in 1968. Although Robinson dealt primarily with steady state noise, the concept has been extended by Atherley et al (ref. 28) and Rice et al (ref. 29) to cover impactive intermittent noise, and gunfire noise, respectively.

30. This work, considered with that of Burns et al (ref. 30) in 1970 made possible the calculation of the risk to hearing represented by various combinations of noise exposure. By means of a large retrospective study, Burns showed that for a given noise exposure pattern the proportion of the employee population at risk could be predicted, together with the eventual extent of the hearing damage.

31. As a result of these investigations, an acceptable limit was proposed by the Code of Practice for exposure to noise, described by the parameter known as the equivalent continuous sound level (L_{eq}, with units of dBA). This quantity provides a method of summating the total A-weighted noise energy over an eight hour working day, to produce a single figure. A continuous noise exposure of 90 dB(A) over eight hours would have an L_{eq} rating of 90 dB(A). An exposure of 93 dB(A) lasting for only four hours in a working day would also have an L_{eq} rating of 90 dB(A). Both stimuli contain the same sound energy, and hence receive the same L_{eq} rating normalized to an eight hour working day. The time intensity trading/relationship is such that for each L_{eq} increase of 3 dB(A), the time of exposure must be halved for the amount of auditory hazard to remain constant.

32. As an upper limit the Code of Practice states that the unprotected ear should not be exposed to levels of 135 dB(A) or above, regardless of the brevity of the time of exposure. As the L_{eq} is a measure of the total energy received during the working day, it follows that intermittent or fluctuating noise exposures can be quantified in terms of an L_{eq} rating. Tables used for carrying out these calculations for intermittent exposures are given in the Code of Practice, and are explained more fully in Burns (ref. 31).

33. The sound level set as being acceptable by the Code of Practice is that with an L_{eq} rating of 90 dB(A) or below. It should be pointed out, however, that this level is still not "safe". Using the formulae given in Burns (ref. 30) it can be shown that exposed to a noise level of 90 dB(A) L_{eq} for a working lifetime (45 years) 22% of the exposed population at age 65 will exhibit a hearing loss in excess of 30 dB averaged over the audiometric frequencies 1, 2 and 3 kHz. The hearing loss, which excludes pathology, comprises of two components: that due to noise (approximately 18 dB) and that due to ageing (presbyacusis, approximately 12 dB). The British Standard method of test for estimating the risk of hearing

handicap due to noise exposure(B.S.5300, (ref: 32)) would define the 22% as the handicap percentage, as hearing handicap is deemed to exist when the level of loss exceeds 30 dB, averaged over 1, 2 and 3 kHz.

HEARING CONSERVATION PROCEDURES

34. In the case of an environment in which the noise exposure is above the recommended level, steps must be taken to ensure that the L_{eq} at the ear of the employee is below the 90 dB(A) limit.

35. A noise survey must be undertaken to identify hazardous areas and machinery. Four approaches to the problem are then possible, and are given in descending order of acceptability.

36. The best procedure is to attack the problem at source, either quietening or replacing the existing machinery. If this is not practically or economically possible, then the possibility must be investigated of changing the work pattern of the individual so that he is no longer exposed to a noise hazard. Thirdly, acoustic barriers can be placed between the offending machinery and the employee, and if this fails, as a final resort the employee must wear personal hearing protection. However, ensuring the proper and continued use of hearing protection can be more difficult than might at first be expected (Karmy, ref. 33). If hearing protectors are issued, then a programme of education must be carried out revolving around the hazard represented by excessive industrial noise, together with instruction on the use and maintenace of hearing protectors, and continuous supervision of the protector programme.

37. It is usually appropriate, at the discretion of the Works Medical Officer, to begin a programme of regular audiometric (hearing) tests for personnel at risk, both to ensure that the hearing protection programme is being successful, and to identify the small percentage of employees who would still become industrially deafened whilst working in a 90 dB(A) L_{eq} environment (Martin, ref. 34).

LEGAL ASPECTS

38. The two main functions of the legal mechanisms related to noise in the United Kingdom are prevention and compensation. Preventative documents include the Factories Act (1961), Noise and the Worker (Department of Employment 1971), The Code of Practice (Department of Employment 1972), Woodworking Machines Regulations (Department of Employment 1974) and the Health and Safety at Work Act (1974). Processes for compensating the noise deafened employee are threefold; the National Insurance (Industrial injuries) Act of 1965, Common Law, and out of court settlements.

39. Section 29(1) of the Factories Act compels any employer to maintain a safe working environment. Although this includes the removal of any noise hazard, in practice no action was ever brought by the Factory Inspectorate against an offending employer. Noise and the Worker, and the Code of Practice are both advisory documents, and as such only have real weight when used in the courts. The Woodworking regulations, on the other hand, do possess legal force, and give effect to the noise limits set out in the Code of Practice. As the name suggests, they only apply

to woodworking machines. The Health and Safety at Work Act (1974) came into effect on April 1st, 1975. Whilst not referring specifically to noise it again compels the employer to make safe the place of work, but in addition it requires the employee to be responsible for complying with any safety procedures laid down by the employer. The Health and Safety at Work Act is an enabling act, that is, it makes possible the swift passage of any future specific regulation concerning noise on to the statute books. The specific regulations are expected in 1977/78 and the format is discussed in a document produced by the Health and Safety Executive (ref. 35), Framing Noise Legislation.

40. Until 1974, the National Insurance Act covered only hearing damage caused by a noise "accident", i.e., a blast or sudden intense noise exposure. In 1974, however, occupational deafness became a prescribed disease, and as such is compensatable by the State under the National Insurance scheme. At the present time however the benefits are limited to employees who have worked for at least 20 years in the shipbuilding, ship repairing or metal manufacturing industries. All claims must be made within one year of leaving the prescribed industry. Compensation is currently only awarded when disability reaches 20%, which in this case represents a hearing loss in the better ear of 50 dB averaged over the audiometric frequencies 1, 2 and 3 kHz. Compensation is then proportional to increasing disablement, reaching 100% disability in a directly comparable way to disability caused by other industrial injuries. The scope of the compensation scheme is currently under review, and is expected to be widened to include other industries in the near future.

41. Common Law can be a much more lucrative vehicle by which to obtain a cash settlement. This involves the employee suing the employer for a noise induced hearing loss sustained during a period of employment. The plaintiff must show that "on the balance of probabilities" it is "more likely than not" that the loss is a noise-induced one sustained as a result of negligence on the part of his employer. Furthermore the damages may be reduced by a certain percentage if the employee is considered to have shown contributory negligence by refusing to comply with safety measures initiated by the employer. As a result of recent large settlements obtained in the compensation courts many claims are now settled out of court.

CONCLUSION

42. The alarmingly high incidence of noise induced deafness among the population of industrial workers represents a definite social problem. As with all the occupational problems faced by industry since the industrial revolution, the solution is expensive. However, in view of the present social and legislative climate pressure on the employer is increasing, and there is no doubt that these problems will have to be overcome during the next two decades.

REFERENCES

1. PLINY, -. A Natural History. I A.D. Cic. Somn. Scip. c.5 vi 35.
2. BACON, F. Sylva Sylvarum. 1627 pub. Rawley, W., London.
3. RAMAZINI, B. De Morbis Artificum. 1713 Padua. (1964 edition Disease of Workers) pub. Hafner, N. York.
4. FOSBROKE, J. Pathology and treatment of deafness. 1831 Lancet 19, p.645.
5. BARR, T. An enquiry into the effects of loud sounds upon the hearing of boilermakers and others who work in noisy surroundings. Proc. Glasgow phil. soc. 17, p.223.
6. COLMAN, B.H. Experimental aspects of reconstructive surgery. The Ear. 1976 in Scientific Foundations of Otolaryngology. ed. Hinchcliffe and Harrison. Pub. Heinmann Medical Books.
7. HUNTER-DUVAR, I.M. and BREDBERG, G. Effects of intense auditory stimulation: hearing losses and inner ear changes in the chinchilla. 1974 J. Acoust. Soc. Am. 55, No. 4, April. p.795-801.
8. BOHNE, B.A. Mechanisms of noise damage in the inner ear. 1976 in Effects of Noise on Hearing. ed. Henderson, Hamerruk, Dosanjh, Mills. Pub. Raven Press, New York.
9. DIEROFF, H.G. and BECK, C. Experimentell-mikroskopische Studien zur Frage der Lokalisation von bleibenden Hörschaden nach Industrielärmbelastung mit tonalen Gerävschanteilen. 1964 Arch. Ohr. Nas. Kehlkopfheilk 184, vol. 33-45.
10. HAWKINS, J.E., JOHNSSON, L,G., STEBBINS, W.C., MOODY, D.B. and COOMBS, S.L. Hearing loss and cochlear pathology in monkeys after noise exposure. 1976. Acta. Otolaryngol. 81, pp. 337-343.
11. HUNTER-DUVAR, I.M., and ELLIOTT, N. Effects of intense auditory stimulation: hearing losses and inner ear changes in the squirrel monkey. 1972. Journal Acoust. Soc. Amer. 52, No. 4 part (2), pp. 1181-1191.
12. SPOENDLIN, H. and BRUN, J.P. Structural alterations due to different types of acoustic trauma. 1976. In Disorders of Auditory Function II. Ed. Stephens, pub. Academic Press, London, pp. 29-36.
13. WARD, W.D. and DUVALL, A.J. Behavioural and ultrastructure correlates of acoustic trauma. 1971. Ann. Otol. Rhinol. Laryngol. 80, pp. 881-896.
14. ELDREDGE, D.H., MILLS, J.H. and BOHNE, B.A. Anatomical behavioural and electrophysiological observations on chinchillas after long exposure to noise. Intern. Symp. Otophysiol. Ed. Karger, pub. White Plains, New York.
15. WARD, W.D. Personal communication. 1976.
16. KIANG, N.Y.S., LIBERMAN, M.C. and LEVINE, R.A. Auditory nerve activity in cats exposed to ototoxic drugs and high intensity sounds. 1976. Ann. Otol. vol. 85, pp. 752-767.
17. JANSEN, G. Zur nervösen Belastung durch Lärm Beihefte zum Zentralblatt für Arbeitsmedizin und Arbeitsschutz. 1976. Heft 9, Darmstadt. Dr. Deitrich Steinkopff. Verlag.
18. PERLMAN, H.B., KIMURA, R. Cochlear blood flow in acoustic trauma. 1962. Acta Otolaryng. 54, pp. 99-110.
19. WARD, W.D. Noise induced hearing damage. 1973. Otolaryngology, vol. 2. Eds. Paparella, M.M., Shumrick, D.A. Pub. Saunders, Philadelphia, pp. 377-390.

20. ENGSTRÖM, H., ENGSTRÖM, B. and ADES, W.
Effects of noise on Corti's organ. 1976. In Man
and Noise, ed. Rossi, Vigone. Pub. Minerva
Medica. Turin. pp.75-91.

21. COLES, R.R.A. Relationships between
noise induced threshold shifts, morphological
change and social handicap. 1975. Symp. Zool.
Soc. London No. 37, pp. 107-120.

22. KELL, R.L., PEARSON, J.C.G., ACTON, W.I.
and TAYLOR, W. Social effects of hearing loss
due to weaving noise. 1971. In Occupational
Hearing Loss, ed. Robinson, D.W. pp. 179-191.

23. JOHNSSON, L. and HAWKINS, J.E. Degeneration
patterns in human ears exposed to noise. 1976.
Ann. Otol. 85, pp.725-739.

24. DEPARTMENT OF EMPLOYMENT A code of practice
for reducing the exposure of employed persons to
noise. 1972 H.M.S.O. London.

25. HEALTH AND SAFETY EXECUTIVE Health and
Safety at Work, etc., Act. 1974 Chapter 37.
H.M.S.O. London.

26. UNITED STATES AIR FORCE Hazardous noise
exposure 1956 Air Force Regulation No. 160-3,
Washington D.C. Department of the Air Force.

27. ROBINSON, D.W. The relationship between
hearing loss and noise exposure. 1968. National
Physical Laboratory Aero. Report AC32, N.P.L.England

28. ATHERLEY, G.R.C. and MARTIN, A.M. Equivalent continuous noise level as a measure of injury
from impact and impulsive noise. 1971. Ann. occup.
Hyg. vol. 14, pp. 11-28.

29. RICE, C.G., and MARTIN, A.M. Impulsive
noise damage risk criteria. 1973. J. Sound Vib.
28, pp. 359-367.

30. BURNS, W. and ROBINSON, D.W. Hearing and
noise in industry. 1970. H.M.S.O. London.

31. BURNS, W. Noise and Man. 1968, pub.
Clowes, London.

32. BRITISH STANDARDS INSTITUTION. Method of
test for estimating the risk of hearing handicap
due to noise exposure. 1976. H.M.S.O. London.

33. KARMY, S.J. Hearing protection. 1977.
In Hearing Conservation. Ed. Bryan, Tempest.
Salford University, in Press.

34. MARTIN, A.M. Industrial hearing conservation: audiometry. 1976. Noise Control,
Vibration and Insulation, March.

35. HEALTH AND SAFETY EXECUTIVE. Framing Noise
Legislation. 1976. H.M.S.O. London.

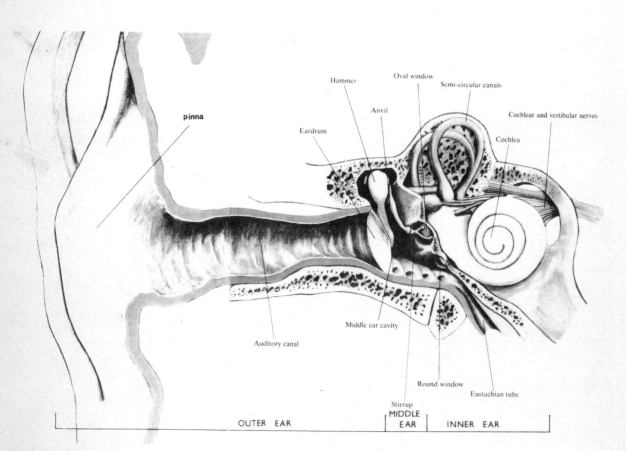

Fig. 1: Longitudinal section through the human ear (Bilsom, Henley)

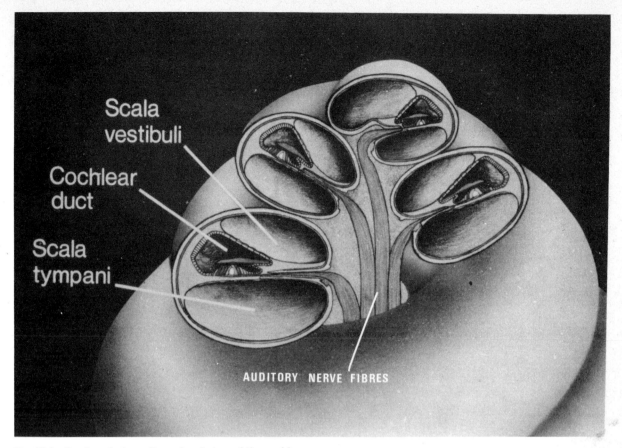

Fig. 2(a): Sectioned, three dimensional view of the cochlea

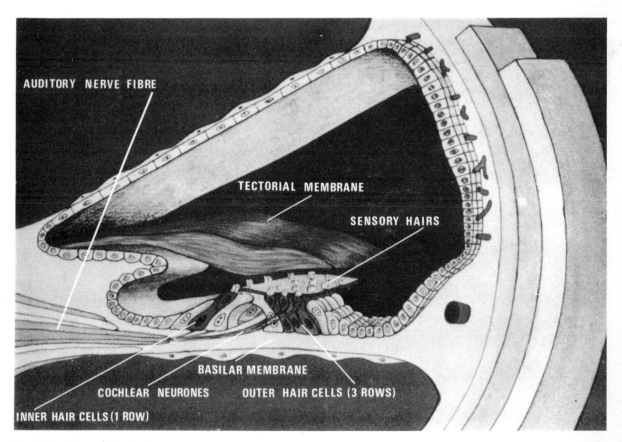

Fig. 2(b): The cochlear duct

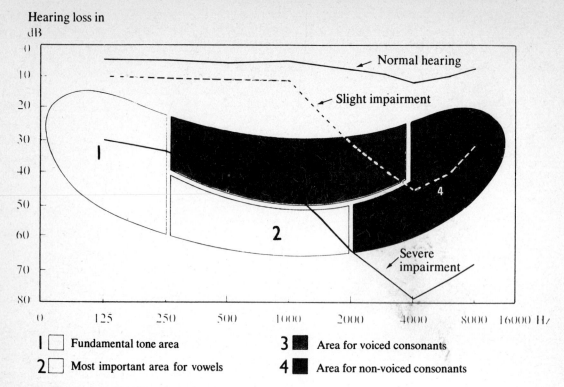

Fig. 3: The progression of a noise induced hearing loss.
The parameter is increasing time of hazardous exposure (Bilsom, Henley)

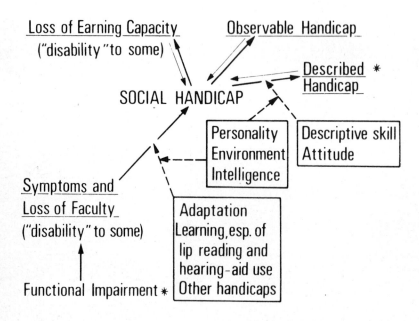

✳ Main measurements in research on social handicap
caused by noise-induced hearing loss.

Fig. 4: Factors influencing the concept of social handicap (Coles, ref 21)

C247/77

THE NEED FOR THE CLASSIFICATION OF PUMP NOISE AND VIBRATIONS

D. FRANCE, BSc, MSc, AFIMA,
Weir Pumps Ltd., Cathcart, Glasgow.

The MS of this paper was received at the Institution on 1st June 1977 and accepted for publication on 19 July 1977

SYNOPSIS

Pump manufacturers are now more than ever being faced with specifications which demand lower noise and vibration limits from their machines.   In many instances, customer standards of acceptance or limits are set which have little relation to the Manufacturer's Codes of Practice and experience. This situation has arisen in default of a collective Pump Manufacturer's document specifying noise and vibration limits for pumps in a manner analogous to Motor Manufacturers.

The paper sets out by reviewing current noise legislation and its import for rotating equipment and goes on to discuss noise and noise control of centrifugal pumps.   The paper considers briefly certain vibration standards which are frequently applied to pumps and argues the need for guidance limits for noise and vibration to be provided by the Manufacturers themselves.

INTRODUCTION

1.   Pump manufacturers are now more than ever being faced with specifications which demand lower noise and vibration limits from their machines.

2.   The publication of documents such as the Health and Safety at Work Act 1974 and the Control of Pollution Act 1974 have implicitly, rather than explicitly, set limits for noise emission from machinery.   A document relating to the reduction of noise from new machinery is awaited but this is unlikely to become law, at least not for some considerable time, since the one document could hardly expect to set limits or guidelines for noise reduction which could be considered appropriate for the whole range of mechanical equipment producing noise.   Noise limits appropriate to sewing machines are far less severe relatively speaking than for, say, high speed boiler feed pumps.

A characteristic of all rotating machinery is, of course, its vibration signature and whilst there is no legislation relating to limits contractual arguments often rage about interpretation and relevance of vibration standards. Many prospective customers or their consultants specify vibration limits which do not necessarily take cognisance of the mounting detail of the pump, the constraints and influences imposed by works test or the manufacturer's own acceptance standards built up by experience of measurement on test and in service.

3.   It is clear that manufacturers of pumps have a responsibility to the community in terms of the intrinsic safety of their machines and it is not the purpose of this paper to challenge the wisdom and desirability of this situation.   Equally, it is not suggested that manufacturers are not willing to make the investment necessary to investigate and apply the methods of reducing noise and vibration at the design stage.   The author's Company has for example, made a considerable investment in the last five years in providing facilities and staff to undertake basic noise and vibration reduction work.

4.   This paper has been written in the first instance to draw the attention of prospective customers and/or noise legislators to the fact that much pumping equipment of modern design and manufacture cannot be expected intrinsically to meet limits of noise as low as 85 dBA. Secondly, it is hoped to illustrate some of the difficulties experienced by pump manufacturers due to customer insistance on strict adherence to vibration guideline documents even in the face of reasoned argument by the manufacturer for a higher vibration norm in specific cases. Thirdly, and I think more important, is the fact that collectively, pump manufacturers have not taken seriously the trend towards tighter control of noise and vibration and this reflects in a total absence of any document specifying noise and vibration limits for pumps in a manner analogous, for example, to the motor manufacturers specifications BEAMA 225 and BS.4999 Pt.51 (Ref. 1 and Ref. 2).

5.   The paper therefore develops its theme by reference to noise data obtained from centrifugal pumps both on test and in service, and suggests some approaches to noise control in view of ever reducing noise limits.   It continues by reviewing some popular vibration standards which are commonly used for pumps. Finally, a plea is made for guideline limits of noise and vibration agreed by the pump manufacturers themselves.

NOISE LEGISLATION FOR ROTATING EQUIPMENT

6.   The limit placed on sound pressure of any

piece of equipment is usually fixed by hearing conservation criteria. At the present the recommended limit is 90 dBA but there is a body of opinion which suggests that 85 dBA should be the maximum full shift habitual exposure. The Health and Safety at Work Act 1974, while not specifically setting limits for noise, does mention that it is the responsibility of the employer to ensure that adequate provision is made to safeguard employees at work. By implication, this means that damage caused by hazardous noise is included.

7.   In Europe things are little different and in West Germany there is a recommended limit of 85 dBA for hearing conservation with 90 dBA as the maximum allowable.

8.   The restriction placed on machinery noise by environmental pressures is covered in the Control of Pollution Act 1974. In the case of an existing plant or the building of a new plant the noise level at the plant boundary due to all sources has to be fixed by negotiation with the Local Planning Authority. This noise level can be traced back to items of individual plant and specified as a sound power output. In many instances the limitation of sound power of the machine implies very low sound pressure at 1 metre from the machine surface.

9.   For instance, suppose a small pump produces a sound pressure at 1 metre from its casing of 90 dBA. Assume the pump dimensions are small in relation to 1 metre then the sound power expressed in dBA re 10^{-12} watts, assuming hemispherical radiation of noise, is:

$$\text{Sound Power} = 90 + 10 \log_{10} 2\pi(1)^2$$
$$= \underline{98 \text{ dBA}}$$

10.   Now, if the sound pressure at the boundary, say 10 metres distance, due to this one pump has to be 45 dBA then the sound power must be limited to, ignoring absorption in the air,

$$\text{Sound Power} = 45 + 10 \log_{10} 2\pi(10)^2$$
$$= \underline{73 \text{ dBA}}$$

i.e., 25 dBA less than its actual sound power. This means the pump would have to be enclosed to comply with the community requirements but not to comply with hearing conservation criteria.

NOISE OUTPUT OF CENTRIFUGAL PUMPS

11.   For the purposes of being able to estimate the noise level of any pump at the design stage a method was evolved at the author's Company and is described in Ref. 3. The chart shown in Fig.1 is reproduced from this reference and indicated on this are the results from some recent noise tests on a pump in an Anechoic Room at Weir Pumps. The free field measurement results support the view that the estimating method is conservative. The reason for the conservatism is because most noise data were obtained from Works tests and so therefore includes noise of the pump and driver and the noise of the pressure breakdown device used to dissipate pump head. The method of measurement used on test, the layout of pipework, contribution from prime mover noise and pressure breakdown device are therefore important in relation to meaningful noise tests on the pump and

this should be an area where manufacturers should agree methods and standards.

12.   In general, centrifugal pumps have tended to become more compact as a result of improved materials, improved manufacturing techniques and NDT techniques and the need to reduce direct costs. Unfortunately, this has had the effect of increasing the noise and vibration of the machines and it is clear from the chart shown in Fig. 1 that pumps of more than 300 kW (at 3000 rpm) or 600 kW (at 1500 rpm) are likely to produce 1 metre noise levels in excess of 90 dB. (On an 'A' weighted scale the noise level is likely to be in the range 85-90 dBA depending on speed.) Clearly, Boiler Feed Pumps with heads per stage in excess of 1100 metres and power/stage at 9000 kW produce 1 metre noise levels in the region of 100-105 dBA which is 20 dBA above the possible future requirement for hearing conservation.

POSSIBLE NOISE CONTROL MEASURES FOR CENTRIFUGAL PUMPS

13.   The airborne noise radiated from centrifugal pumps can be reduced in two basic ways:

1.   Reduction of noise at source.

2.   Absorption or containment of noise by external means.

14.   Reduction of noise at source is favoured as the more elegant of the two but this route is not necessarily the most cost effective way of achieving the required noise reduction.

15.   The noise generated by centrifugal pumping action falls into four main groups.

16.   The first is interference noise resulting from the interaction between the flow at entry and discharge of the impellers and the rotating and stationary components respectively. The second is turbulence noise resulting from vortices set up in the flow. The third is cavitation noise resulting from the collapse of the vapour bubbles produced when the local static pressure falls below the vapour pressure. The last is noise due to mechanical sources.

17.   In general, many of the design features which affect one source of noise will also affect another source in a similar way, e.g. a badly designed flow path which induces turbulence is also likely to induce high local velocities and hence low static pressures which would imply the risk of cavitation.

18.   The choice of operating condition is important and the airborne noise is least in the region of the best efficiency point on the pump characteristic. When operating at best efficiency the outlet flow angle from the impeller should match the diffuser blade angle. Experience suggests that airborne noise reductions of up to 8 dB can be achieved by matching the required duty to the best efficiency duty of the pump. The octave band spectra shown in Fig. 2 demonstrates this as does the results of free field measurements shown in Fig. 1.

19.   Even when the pump is operating near to its best efficiency, reductions in the airborne noise can be made by reducing the interference

between the rotating or stationary flow actions. Thus any inlet guide vanes should be as far away from the impeller as possible without impairing the intended flow straightening. The wakes resulting from the blading, both stationary and rotating, should be as thin as possible. This can be achieved by contouring the blades to produce an even pressure distribution over the blade chord, and by keeping blade trailing edges thin. Poor inlet flow conditions whether resulting from contorted pipework, obstructions or indifferent sump conditions can increase the airborne noise generated by all four mechanisms.

20. Because of its high frequency content cavitation noise is the most unpleasant and damaging form of noise produced by centrifugal pumps. When considering cavitation noise (and turbulence noise) the design of the pipework and valves is just as important as the design of the pump itself.

21. The 'mechanical' sources most likely with pumps are due to rolling element bearings and drive line couplings, although high speed couplings can be a source of aerodynamic noise. The greatest noise reduction, however, is likely to be achieved by reducing the blade loading, i.e. by larger diameter impellers generating less head. This is the most uneconomic solution as it results in larger, slower speed and thus more expensive pumps.

22. Where noise reduction at source is not possible or economic, then absorption and containment of noise must be used for control.

22. The radiated noise may be reduced, where appropriate by vibration isolation, the use of sound deadened materials, surface treatment and acoustic enclosure. The use of sound deadened materials and damping treatments are only effective on lightweight panels such as coupling guards etc. but these often represent the worst noise radiators and the simplest cures. Further reductions in radiated noise can be achieved by cladding the pump. The form of the close cladding can be affected by the need for thermal lagging, but the best combination seems to be a combination of a relatively dense inner layer of say fibreglass or ceramic thermal lagging combined with a less dense outer sound barrier. The use of cladding and pump design must be interlinked as the cladding can only be fully effective with total covering including the pump supports and any appendages. According to Ref. 4 reductions of between 3 to 15 dB are possible by cladding. Finally, acoustic enclosures of good design can achieve very high noise reductions of the order of 30-40 dBA but they can be expensive and difficult to fit to pumps due to secondary pipework.

VIBRATION STANDARDS FOR CENTRIFUGAL PUMPS

23. To the author's knowledge there is only one vibration standard which is applicable to centrifugal pumps, API 610. In this standard limits are given for shaft relative to bearing vibrations for sleeve bearing pumps and absolute casing vibration for anti-friction bearings. The limitations of this standard seem to be:

1. It applies to end suction volute pumps with overhung impellers where shaft orbits at the bearings give a good measure of

impeller deflection.

2. It applies to shaft diameters up to approximately 3" but shaft relative to bearing vibrations are not related to bearing clearance.

24. The use of this standard for multistage diffuser pumps is common but in the author's opinion unduly restrictive because:

(i)   Diffuser pumps produce smaller radial hydraulic loads than volute pumps and hence the bearing specific loads are less.

(ii)   Smaller specific loads and more generous bearing clearances give rise to larger shaft orbits at the bearings.

(iii) The size of shaft orbits at the bearings give no indication of the impeller deflections.

25. In contract machinery vibration guidelines such as VDI 2056 and BS.4675 are based essentially on the Rathbone Curves and assessment criteria are based on the vibration velocity measured on the bearing housing as the critical quantity. The vibration intensity of harmonic vibrations is expressed through the rms value of vibration velocity.

26. In the author's experience, many pumps exhibit vibration levels which do fall within the guidelines laid down in these publications. Equally well, there are many instances of pumps where vibrations fall outside these. This is particularly so with vibration occurring at part load, vibrations occurring at impeller blade passing frequency and vibrations of machines on test on temporary support structures. One of the more annoying recurrences in the author's experience is the rigid adherence of purchasers to the same standard for vertical machinery as for horizontally mounted machines of the same speed and power. The proven reliability of such pumps in service demonstrates the nonsense of inflexible attitudes often demonstrated by purchasers in respect of vibration limits.

CLASSIFICATION OF NOISE AND VIBRATIONS

26. Some of the arguments and evidence so far presented in this paper serve to illustrate the difficulty pump manufacturers will have in the future in meeting noise and vibration limits specified to them. It is hoped that some of the statements made in relation to noise expectations for pumps in excess of 300 kW will be accepted by users and comments made about vibrations.

27. Unlike the motor manufacturers, pump manufacturers have not yet presented information on realistic and achievable noise and vibration levels consistent with reliable operation. In the absence of such data purchasers cannot be blamed for the inflexible attitude they sometimes show in the application and interpretation of standards which in many cases are inappropriate to the pumps under consideration.

28. It is suggested therefore that pump

manufacturers could make a valuable contribution to the work of noise legislators or standards organisations involved in preparation of rotating machinery vibration guidelines if they were prepared to undertake the preparation of a document giving guidance limits of noise and vibration for a wide range of centrifugal pumps.

29. The author has given this some considerable thought and is prepared to venture his opinion on the scope of such a document.

30. The document would need to:

1. Specify measuring positions, measurement standards and operating conditions.

2. Make a statement drawing attention to the difficulties often experienced when testing a pump in the manufacturer's works. It would emphasize those features which affect noise for example, valve noise interference generated by head breakdown. For vibration it would need to stress the difference between pumps which are tested and or supported in a similar manner to the service conditions and those pumps tested in plainly dissimilar operating and/or support conditions from those to be met in service.

3. Specify limits of noise for new pumps tested in either free field conditions or under the conditions typical of manufacturer's works.

4. Specify limits of vibration severity for new pumps obtained from test in the manufacturers works and classified according to whether the test conditions etc. are similar or dissimilar to the service conditions.

5. Make a statement with regard to rejection of firm acceptability criteria for vertical pump vibrations tested in the manufacturer's works and include data from tests on some units to demonstrate the order of magnitude of vibrations and the possible variations.

6. Produce guidance notes on the possible modes of operation of pumps to ensure least noise and indicate some noise control methods for pumps.

31. A document of this type would be a very useful addition to the information that exists for other types of rotating equipment and could help for instance, prospective purchasers in planning noise control of complete plants before actually going out to tender.

CONCLUSIONS

32. It is stated that it will not be possible for pumps above about 300 kW (at 3000 rpm) to meet 1 metre sound pressure levels of less than 85 dBA.

33. The limits of noise placed upon us by environmental considerations will not be met by intrinsically quieter machines but by machinery which is suitably enclosed or housed to interrupt the path of noise propagation. Small reductions in noise from pumps is possible by good design and by attention to the mode of operation, the pipework layout and suction conditions. Greater reductions at source can be made but the penalty in terms of size and cost seems prohibitive.

34. Vibrations in pumps are complex and existing machinery vibration standards are not always applicable. In particular, noise and vibrations obtained from works test are likely to be higher than obtained at site in the service condition.

35. It is recommended that pump manufacturers should prepare a guideline document relating to noise and vibration limits of a wide range of pumps stressing the difficulties of testing in works and specifying conditions and methods by which useful measurements should be made.

ACKNOWLEDGEMENTS

The author would like to thank colleagues for assistance in the preparation of this paper and Weir Pumps Ltd. for permission to publish.

REFERENCES

1. BEAMA 225

2. BS.4999 Pt. 51

3. Some factors in the measurement and estimation of airborne noise from centrifugal pumps. I.Y. Reid. I.Mech.E. Conference on Vibrations and Noise in Pumps, Fans and Compressors 1975.

4. Techniques for Noise Exposure Control in Existing Power Plants. R.A. Popeck. Journal of Engineering for Power. October 1976.

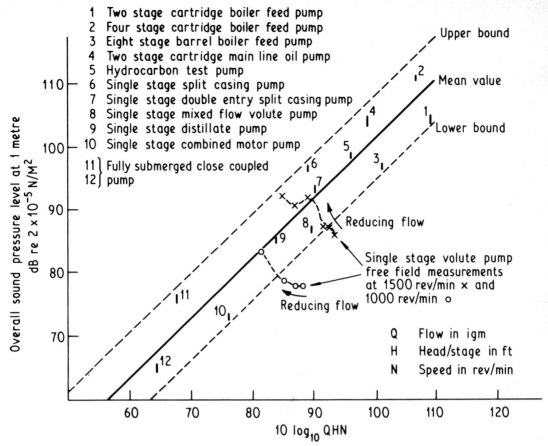

1 Two stage cartridge boiler feed pump
2 Four stage cartridge boiler feed pump
3 Eight stage barrel boiler feed pump
4 Two stage cartridge main line oil pump
5 Hydrocarbon test pump
6 Single stage split casing pump
7 Single stage double entry split casing pump
8 Single stage mixed flow volute pump
9 Single stage distillate pump
10 Single stage combined motor pump

11 ⎫ Fully submerged close coupled
12 ⎭ pump

Upper bound
Mean value
Lower bound

Reducing flow

Single stage volute pump
free field measurements
at 1500 rev/min × and
1000 rev/min ○

Reducing flow

Q   Flow in igm
H   Head/stage in ft
N   Speed in rev/min

Fig. 1: Curve used for estimating overall noise level of a pump

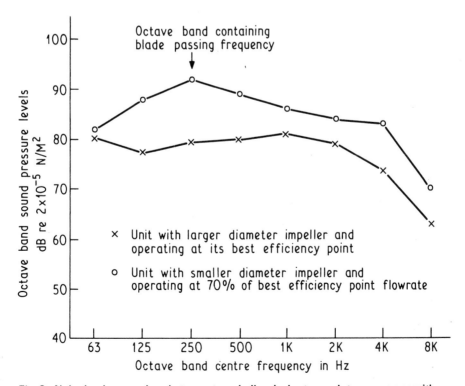

Octave band containing
blade passing frequency

× Unit with larger diameter impeller and
   operating at its best efficiency point

○ Unit with smaller diameter impeller and
   operating at 70% of best efficiency point flowrate

Fig. 2: Noise level comparison between two similar single stage volute pumps one with
a much reduced impeller diameter and operating at 70% of its best efficiency
flowrate

13

C248/77

DEVELOPMENT OF A STANDARD, SILENCED, OIL INJECTED, SCREW TYPE, AIR COMPRESSOR PLANT

G.L. COLLIER, CEng, MIMechE
CompAir Industrial Limited, High Wycombe, Bucks.

The MS of this paper was received at the Institution on 14th July 1977 and accepted for publication on 1st August 1977

SYNOPSIS   The sources of noise from an enclosed, air-cooled, screw compressor plant are distinguished as airborne, such as cooling fan noise, and structure generated resulting from the compressor delivery pressure pulses. Systematic noise testing of an enclosed plant with various modifications based on in-situ experimental work, and tests on cooling air intake and exhaust systems, etc, in an acoustic laboratory, a noise level of 70 dB(A) was obtained for the plant.

INTRODUCTION

1.   The totally enclosed screw compressor plant is now well established in the air compressor market. This type of equipment (illustrated by a sectioned sketch in Figure 1) offers the user virtually all the ancillary equipment he requires for a complete compressed air supply installation and by offering such equipment the manufacturer tacitly accepts the responsibility for the selection of all the component parts of the installation. These include the pressure vessels, the coolers, electrical equipment, protective instrumentation and the sound attenuating enclosure which houses these items. As this type of equipment is becoming more firmly established so silencing requirements are becoming more stringent and the manufacturers of screw type air compressor plant must be able to offer plants which meet these requirements.

2.   Any manufacturer attempting to design and supply air compressors of this type to a wide international market, must ensure that the machinery is suitable for a wide range of operating conditions and duties. Typically, he will offer several machines of broadly similar design and construction covering a range of capacities, each capable of operating at a variety of delivery pressures. The ambient temperature in which the compressors will operate can vary from -40°C to $+45^{\circ}$C and they must be capable of adequate cooling at the maximum ambient temperatures if the necessary long service lives are to be achieved. The noise levels of enclosed and acoustically treated machines are now typically 68-73 dB(A) when assessed as the average of measurements taken at several points around the plant at one metre distance.

3.   This paper describes briefly the approach used to achieve a 70 dB(A) noise level with an air-cooled screw compressor plant designed to operate in the full range of world-wide markets described in the preceding paragraph.

SPECIFICATIONS

4.   The marketing specification presented to the Engineering Department defined the parameters for meeting the requirements of the very different markets throughout the world. Where legislation existed for pressure vessels, electrical equipment, etc, the codes and laws for each country were examined to ensure compliance with their requirements. The specification of 70 dB(A) or less for the noise level was seen as exceeding current legislation and being in line with the downward trend that is expected to continue within the screw compressor market during the next few years .

DESCRIPTION OF THE PLANT

5.   Referring to Figure 1, the screw compressor is of the single-stage, oil injected type consisting of a pair of helically engaged intermeshing rotors housed within a stator casing. The male rotor has four lobes and the female rotor six flutes. The drive is transmitted through a pair of straight cut helical gears to the male rotor, which drives the female rotor. The screw compressor is connected by a flexible coupling to a 55 kW electric motor rotating at 49.1 1/s. The speed of rotation of the male rotor depends upon the gear ratio; this particular machine the male rotor speed is 91.7 1/s. A bell housing rigidly connects together the electric motor and the compressor.

6.   The screw compressor is a positive displacement machine. As the rotors turn, air at atmospheric pressure enters the intake and fills the space between the casing, the lobes of the male rotor and the flutes of the female rotor. Here the air is trapped, compression starts and oil is injected into the air. Continued rotation of the rotors reduces the space for the air/oil mixture until it is compressed to the final pressure. At this point the delivery port in the casing is uncovered and the air/oil mixture passes through a non-return valve to the oil reclaimer.

7.   Within the reclaimer the oil is separated from the air and is pressure forced through the oil cooler and other components to the oil gallery in the screw compressor.

8.   There is no oil pump, oil flow is achieved by the pressure difference in the reclaimer and the atmospheric pressure in the compressor, oil flow rate is controlled by the orifice size in the stator casing.

9.   The compressor unit is isolated from the base frame by three rubber anti-vibration mounts and is coupled to the oil reclaimer and other components with flexible connections.

10.   The combined air blast oil and aftercooler is of the finned tube construction. The 760mm diameter cooling fan is a single-stage, axial type driven by a 24.2 1/s electric motor and capable of producing $3.77 m^3/s$ free air at a fan static pressure of 41mm of water.

11.   The design and construction of the enclosure was based on the ability to manufacture in-house sheet metal components.

12. . Mineral wool covered by mild steel perforated sheet was used for the acoustic material.

DESIGN CONSIDERATIONS

13.   Noise measured around a plant in Free Field conditions can occur as airborne noise and structure generated noise.

Airborne Noise

14.   This radiates from the intake and exhaust ducts of the enclosure and it is transmitted through the components of the enclosure. The noise is generated mainly by the compressor/motor unit, oil cooler fan and other components within the enclosure which are set in vibration by the two prime sources.

15.   With air cooled plants the fan noise is often the major noise source, especially if low noise levels around the plant are sought. The choice of fan is determined by the type of cooler. For example, a high density tube arrangement could be matched by a single-stage axial fan, a guided vane type or even a two-stage axial fan.

16.   During the design study the choice of fan was reduced to two single-stage axial fans (i) 760 mm diameter, 24.2 1/s 16° blade angle, (ii) 535 mm diameter, 48 1/s, 18° blade angle. The attraction of the smaller fan was its size for siting it within the enclosure and because of a fastic tip speed of 80.8 m/s to operate against a higher back pressure. The design point for the original oil cooler was $3.77 m3/s$, total plant pressure drop was estimated at 42mm water. The curves in Figure 2 show that at this design point the 535mm fan should radiate 1 dBW less acoustical power. In practice because there was no sizeable expansion chamber between the fan and cooler, the air passed through a small face area of the cooler, this resulted in a higher back pressure and increased fan noise.

The fan curves are based on smooth flow test conditions in a duct; in the non-ideal conditions that exist within the enclosure the approach air to the fan will not be smooth and the air flows may be lower than expected. Similarly, noise levels may be higher because of the condition of the approach air. One other disadvantage of the smaller fan was that at the design point it absorbed 0.3 kW more than the larger 760 mm fan.

17.   The cooling air path through the plant should ideally be designed with a minimum pressure drop between the fan and the inlet and outlet openings. Unfortunately this opposes noise limiting designs as any significant noise reduction is directly coupled to pressure drop. There is, therefore, a balance between the allowable pressure drop and its effect on the cooling fan and the level of noise reduction.

18.   From considering the cooling air path through the plant in Figure 1 the pressure drops in Table were estimated for an air flow of $3.77 m^3/s$.

Table 1

Estimated cooling air pressure drops.

Across the air intake pre-filter	2.5 mm water
Around the intake baffle	7.6 mm water
Path to the cooling fan	1.3 mm water
Across cooler	20.3 mm water
Across exhaust baffle	8.6 mm water
Leaving exhaust duct	1.5 mm water
	41.8 mm water

19.   Measurements to confirm the values were to be made on the plant using appropriate instrumentation and taking multiple readings.

Structure generated noise

20.   This can be attributed to airborne noise exciting parts of the structure but in the main it is the result of vibration paths linking main noise sources to the structure.

21.   The two main sources are the screw compressor unit and the cooling fan.

22.   The fundamental frequency for this particular compressor is around 365 Hz and is derived from the number of delivery pulses at the compressor delivery port, and from the intake depression variations at the intake port.

23.   The effect of the intake is not usually very pronounced on the structure as the pressure amplitudes are of the order of 3.5 kPa peak to peak and the connection to the intake filter is by a flexible lightweight hose. The airborne pulse is normally damped effectively by the intake filter.

24. The main forcing frequency is that from the pressurized air/oil mixture being discharged through the delivery port. The meshing of the four male rotor lobes with the six female flutes results in a repeatable pattern of pressure variations. The pulses are not equal, this is due to very small differences in rotor profile clearances. Figure 3 illustrates this.

25. The screw compressor has a built-in pressure ratio, this is expressed as the ratio of the space for filling by intake air to the space filled by the compressed air/oil mixture prior to discharge into the delivery port. With intake pressure of 101.3 kPa the internal pressure will be 827.4 kPa. If the delivery port pressure is higher than this value then additional compression takes place within the rotors and delivery port when the mixture is discharged. Similarly, if the delivery port pressure is lower, then rapid expansion takes place. During the compression process within the compressor there is a pressure ripple due to the initial intake pulse, this tends to become damped towards the end of the cycle but is re-energized when it collides with the delivery port condition. Figure 4 illustrates this. Harmonics of the fundamental are present and it is these pulsations which are the main disturbing forces.

26. Close scan noise measurements at the compressor delivery port shows an increase of 2 dB with decreasing delivery pressure, see Figure 5.

27. Immediately downstream of the compressor it is usual to fit a non-return valve to prevent the air/oil mixture exhausting itself through the screw compressor when the plant is shut down. It is practice to fit this directly to the compressor delivery flange or at a convenient position in the pipe run to the oil reclaimer. The type of valve and its siting in the pipework can be detrimental to both the amplitude and frequency of the pulsations and, therefore, the noise emitted from the air end. Reflected path lengths between the valve and delivery port of quarter wavelengths of the fundamental frequency should be avoided.

28. Figure 6 shows how the pulsations can be magnified by fitting the valve direct to the compressor; the particular screw compressor is much larger than the one considered in this paper but this test from a section of earlier work is included to demonstrate the preceding statements. The oil flows are high, simulating tropical or high ambient conditions, and do aggravate the pulsation levels. Harmonics at twice the fundamental are present. Repositioning the valve along the pipework had the effect of reducing the mean pressure by 42 kPa and diminishing the amplitudes by over 50%. Perhaps it should be pointed out that not all designs where the non-return valve is fitted directly to the compressor will have the same kind of effect.

29. The functions of the delivery hose are two-fold - to absorb mechanical vibrations of the compressor and to a lesser extent the pressure fluctuations of the air/oil mixture as it passes through the hose to the oil reclaimer. The lateral and vertical movement of the compressor should be contained by the anti-vibration mounts, if this is not the case then the vibration must be damped by the hose or transmitted to the reclaimer. The natural frequency of the latter is high but it can still be excited to radiate noise.

There is also a direct path to the baseframe through the feet of the reclaimer. The vertical components of the compressor movement should be absorbed by the mounts - a 90% isolation is aimed for in the choice of mount. A correlation has been found to exist between the degree of isolation and direct noise (measured by close scanning) of noise radiating from the baseframe.

30. The design of the fan mounting, if separate from the air end unit, also requires thought as fan vibrations can contribute to structure noise.

EXPERIMENTAL WORK

31. This was approached in two ways - carrying out noise tests and appropriate modifications on the enclosed plant, and parallel with this the testing of component assemblies in an acoustic laboratory. The latter approach was decided as in order to check the effectiveness of certain parts of the enclosure, such as the intake and exhaust duct designs it was considered not practical to isolate them on the plant as repeatable and accurate measurements are required to make meaningful comparisons between designs.

32. It was therefore arranged that specific parts be tested under simulated conditions in an acoustic laboratory.

33. Both approaches are described in paragraphs 34 to 45.

Intake noise

34. A number of full size cooling air intake designs were constructed and tested by the two room method of generating noise in one room and measuring the noise reduction in the other room (Ref.1). Noise reduction values by this method are repeatable with an accuracy of ±0.5 dB.

35. Separate air flow tests were carried out by blowing 3.3 m^3/s cooling air through each intake and measuring their respective pressure drops.

36. Noise reduction values and pressure drops for three intake designs are shown in Figure 7.

37. The shape of the spectrum was as expected with the sound reduction increasing with frequency. The magnitude of sound reduction is related, in a simple case, to the surface area and absorption properties of the materials used and the path that the intake air is forced to take.

38. The pressure drops across the intake systems were slightly higher than expected so that the final design for production was the result of a number of modifications.

39. As the choice of absorbing material is extremely important some of the different materials were checked in the laboratory (Ref.2). The results are shown in Table 2

Table 2

Noise absorption coefficients (α) for various absorbing materials.

Material	Frequency (Hz)							
	63	125	250	500	1K	2K	4K	8K
A	0	.2	.4	.65	.7	.8	.8	1
B	0	.15	.5	.75	.85	.9	.9	1
C	0	0	.15	.4	.65	.75	.9	1
D	0	0	.2	.55	.6	.45	.35	.3
E	.2	.15	.45	.8	.4	.85	.7	0

Key

A	50mm mineral wool/perforated mild steel sheet
B	50mm polyether foam (unskinned)
C	25mm polyether foam (unskinned)
D	12mm polyether foam (continuous skin)
E	25mm polyether foam (continuous skin)

40. An interesting finding was that the 25mm continuous skinned foam had coefficient values approaching the thicker materials, expressed as absorption/cost the material has an obvious application where large amounts of foam are used.

Exhaust noise

41. The combined oil and aftercooler unit with the axial fan fitted was built into a common wall between two rooms. The exhaust duct was simulated and different designs for reducing the airborne noise were constructed. The flow through the fan was adjusted to 3.3m^3/s so as to compare the different designs under realistic conditions. Reduction in sound power levels were used as a basis of comparison (Ref.3)

42. The results shown in Figure 8 confirmed that the designs were acceptable for attenuating the fan noise, they were, however, more constricting than hoped for, so modifications were made to reduce the pressure drop for the production version.

43. By the same two-room method the effectiveness of the enclosure doors was checked. An enclosure door frame was built into the dividing wall and a door fastened in its correct position, different acoustic materials were systematically built into the door and the transmission loss measured. Typical materials tested varied from mineral wool covered with perforated mild steel sheet to polyether foam skinned with PVC.

44. Close attention was paid to the door seal and the closed position of the door. The total length of seal fitted to many types of rotary screw compressor plant is usually long, as the number of doors and removable panels is dictated by the need for service and component accessibility.

Tests were carried out with different seals and the seal deflection necessary for optimum performance. At the lower frequencies there was no appreciable difference but mid-band onwards variations of 2-3 dB were measured.

45. This means that a simple part like a door seal is important as the total seal length, for this machine about 30m, can represent a high leakage path.

Experimental work on enclosed plant

46. All noise measurements were generally taken in accordance with Pneurop CAGI test code (Ref.4) i.e. four positions around the plant, 1m from the surface, 1.5m above the ground, exception being the close scan measurements which were taken within 0.3m from the part being measured.

47. All measuring equipment was of Brüel & Kjaer manufacture.

48. An early build of a prototype plant was with rigid pipe connections between the compressor and other components. Noise measurements around the plant resulted in an average sound pressure level of 77 dB(A).

49. A typical spectrum is shown in Figure 9.

50. Adding mineral wool to parts of the enclosure did not reduce the level significantly and confirmed that the structure generated noise was the dominant noise component.

51. Modifications to de-couple the compressor were introduced, namely the fitting of flexible hoses, etc.

52. The average sound pressure level was reduced significantly to around 73 dB(A).

53. A third octave analysis showed that the fundamental pulse and its harmonics was still significant and that the fan noise was also now a significant contribution to the noise levels.

54. This was confirmed by taking measurements with only the fan running and similarly with the compressor operating and the fan switched off.

55. Vibration measurements were taken and related to close scan noise measurements of the compressor components, the oil reclaimer and enclosure to establish what pattern of mechanical movement was present in order to trim critical components.

56. One significant adjustment was to stiffen the anti-vibration mounts laterally to contain the compressor movement in that direction by the mounts and not by the hose.

57. As a further check on a low structure level the adding of extra insulation within the enclosure resulted in a small measurable difference of noise level around the plant. (Sometimes a modification to reduce airborne noise may be judged as ineffectual because any reduction is masked by structure noise). The sound pressure level was assessed as 72 dB(A).

58. At this relatively low level a subjective assessment of the noise identified one discrete tone which could not be identified on 1/3 octave band analyser, a narrow band analyser would have probably indicated it. On the assumption that the tone was related to the fundamental delivery pulse a small silencer was made to dampen the 365 Hz pulse and its harmonics. It was fitted downstream of the non-return valve. The improvement subjectively was immediately noticeable in that only broad band noise was heard. Practical difficulties in scavenging oil from the silencer did not make the fitting of it a feasible production proposition.

59. Finally, a prototype production plant was built incorporating designs based on the test work described in this section and in the acoustic laboratory tests described earlier. The mean sound pressure value was measured at 70 dB(A), a typical sound pressure trace is shown in Figure 9.

CONCLUSIONS

60. To achieve noise levels of 70 dB(A) or less for enclosed, air cooled, screw compressor plants requires a logical step by step approach and continual re-checking if the development work is to be carried out on a plant.

61. Where possible, certain components, particularly the enclosure cooling air intake and exhaust ducts, should be tested separately as component parts in an acoustic laboratory to measure their true performances.

62. The main sources of noise for this type of compressor plant were found to be:

(a) Fan noise radiating as airborne noise mainly through the cooling air intake and exhaust ducts but also through parts of the enclosure.

(b) Structure generated noise resulting predominantly from the pressure discharge pulses of the compressor and radiated as airborne noise within the enclosure or from the enclosure itself.

63. The following design features should be considered separately to minimize the generating or transmitting of noise.

(a) The characteristics of the cooling air fan.

(b) The required level of isolation efficiency for the compressor unit.

(c) The type of compressor discharge pipe to ensure that the vibrations of the compressor are not transmitted along it to other components.

(d) Position of components, such as non-return valve, likely to influence the compressor delivery port pressure fluctuations.

(e) Design of cooling air intake and exhaust ducts.

(f) Choice of acoustic materials.

ACKNOWLEDGEMENTS

Mr. D.R. Winterbottom of Sound Research Laboratories, Sudbury, for the directing and control of the experimental work in the acoustic laboratory.

APPENDIX

REFERENCES

1. British Standards Institution B.S. 2750:1956 Recommendations for field and laboratory measurement of airborne and impact sound transmission in buildings.

2. British Standards Institution B.S. 3638:1963 Method for the measurement of sound absorption coefficients (ISO) in a reverberation room.

3. British Standards Institution B.S. 848:1966 Fan noise testing.

4. Pneurop – C.A.G.I. Test code for the measurement of sound from pneumatic equipment.

Fig. 1: Sectional arrangement of plant

760 mm Dia. Fan. 24·2 l/s

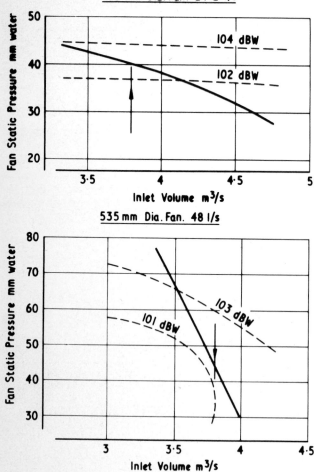

104 dBW

102 dBW

535 mm Dia. Fan. 48 l/s

103 dBW

101 dBW

Fig. 2: Characteristics of two axial fans

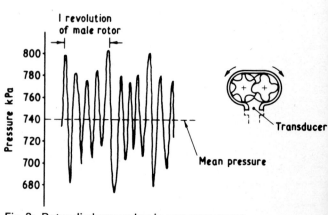

1 revolution of male rotor

Mean pressure

Transducer

Fig. 3: Rotor discharge pulses in compressor port

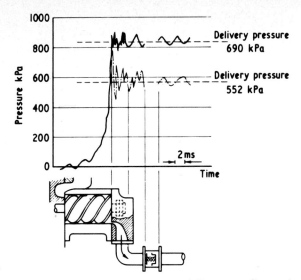

Fig. 4: Effect of discharge pressure on delivery port fluctuations

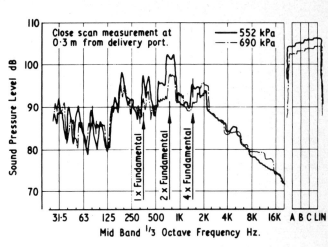

Fig. 5: Effect of discharge pressure on sound pressure level

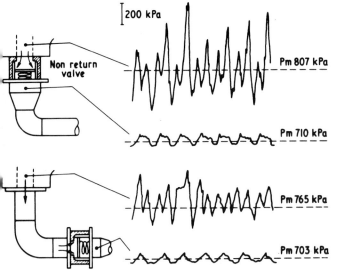

Fig. 6: Effect of non return valve position upon compressor delivery port pulsations

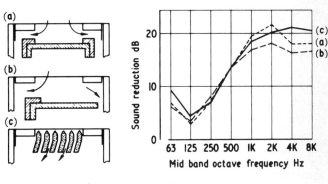

(a) Mineral wool / perf. M.S. P.D. 16·5 mm water
(b) Mineral wool / perf. M.S. P.D. 15·2 mm water
(c) Polyether foam   P.D. 20·3 mm water

Fig. 7: Sound reduction of different air intake ducts

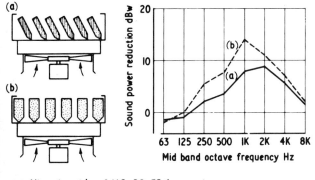

(a) Mineral wool / perf. M.S. P.D. 35·6 mm water
(b) Polyether foam   P.D. 30·5 mm water

Fig. 8: Sound reduction of two designs of exhaust ducts

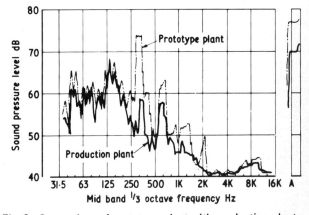

Fig. 9: Comparison of prototype plant with production plant

C249/77

A USER'S VIEW ON THE CONTROL OF NOISE FROM PUMPS, FANS AND COMPRESSORS

J.B. ERSKINE, BSc, CEng, MIMechE, MIEE, FIOA
Noise and Vibration Section, Agricultural Division, Imperial Chemical Industries Limited

The MS of this paper was received at the Institution on 23 June 1977 and accepted for publication on 28 July 1977

SYNOPSIS   Over the last ten years, noise (i.e. unwanted sound) has been the subject of increasing pressure from both the hazard to hearing and nuisance points of view. Chemical Plants have not escaped this pressure and the recent advent of both the Control of Pollution and the Health and Safety at Work Acts have served to sharpen the pressure.

To produce a satisfactory end result for new plant requires that full consideration be given to noise early in the design stage so that equipment can be specified correctly. Beyond the purchase of individual items of plant there is a major need to consider the systems and how they affect the noise performance of a plant. From this, the need to integrate Noise Engineering into a Project Design team is illustrated. This paper explores factors affecting Pump, Fan, and Compressor installations and draws on practical experience to put them into a numerative context.

INTRODUCTION

1. Chemical Plants are major users of Compressors, Fans and Pumps of a wide variety of types, sizes and powers. From this spectrum it is inevitable that there are problems which arise in the control of noise levels to satisfy requirements imposed by documents such as the Department of Employment Code of Practice for the protection of hearing of employed persons, BS 5228 relating to noise from construction sites, and BS 4142 relating to noise in residential areas. The various pressures to limit noise levels have grown significantly over the last ten years and have inevitably led to more detailed consideration of noise in the design of new plant, apart from the need to improve some existing equipment.

2. Implementation of satisfactory and economic noise control measures can best be done during the design stage and the remainder of this paper is intended to show how one user sets about this task. Cooperation between the User, Design Contractor and Equipment Vendors is vital to the achievement of noise targets and some consideration is given both to gaps in existing knowledge and points at which failure to communicate can lead to a poor end result.

ALLOCATION OF NOISE LEVELS TO EQUIPMENT IN A PLANT COMPLEX

3. Because one must consider both hazard to hearing and nuisance it is not possible to simply say that the noise levels around the various items of plant shall not exceed the level beyond which hazard to hearing will occur. Consideration must be given to both aspects; in particular the generation of pure tones must be considered for noise nuisance.

4. The procedure followed by ourselves is to work back from the perimeter sound pressure levels which have been defined, and normally agreed with the Local Authority, as being acceptable from a nuisance to local residents point of view to the permissible sound powers each major area of plant should emit. This is then put alongside the requirements for hearing hazard and nuisance considerations in Control Rooms and nearby office blocks. From these considerations a general noise specification emerges, a typical example is given in Appendix 1: specifications for individual plant items are evolved at the next stage, when detailed Flow Sheets are available, which in total satisfy the various parts of the general noise specification.

FAN NOISE

Prediction of noise levels

5. At the onset of a project only the Fan Duty will be known and recourse has to be made to a simple method of defining sound power. The method used by the author is outlined in depth in Ref 1: essentially one considers only the Fan Power and probable tip speed as a function of Mach No. in the following equation:

$$L_w = K_1 + 91 + 10 \log_{10} kW + 30 \log_{10} \frac{M}{M_{20}} \qquad (1)$$

where

L_w = Sound Power in dB re 10^{-12} Watts

kW = Fan Absorbed Power, kW

K_1 = Octave band correction tabled below

M = Mach No. at gas temperature

M_{20} = Mach No. at 20°C

Octave Band Centre Frequency Hz	63	125	250	500	1000	2000	4000	8000
K_1	-4	-5	-6	-9	-12	-16	-20	-24

This gives a reasonable first order indication of the unweighted sound power due to the Fan alone: noise due to the driving unit must also be considered.

6. From this the need, if any, for a silencer can be defined in the light of decisions made as described in Para 4. Because large fans require large silencers one can carry out an outline design for the purposes of sizing and so setting up equipment layouts to the best advantage: Fig 1 illustrates the size of an inlet silencer on a large FD Fan on a Reformer. It should be noted that this technique does not show up significant pure tone effects: generally, it is prudent to add 5 dB to the octave which contains the blade passage frequency. Nuisance from pure tones frequently arises from small fans which emit directly to atmosphere at a height, e.g. ID fans on ventilation systems or dedusting systems and experience has shown one should not ignore them on the grounds of small size and power.

Control of fan noise

7. Pressure drop limitations very often preclude the use of anything other than absorption type silencers in the fan suction or delivery: a typical arrangement is shown diagrammatically in Fig 2. If it is required **significantly to attenuate frequencies below 500 Hz** the thicknesses of the splitters must be greater than 200 mm and the overall silencer dimensions therefore become large. For further detail the reader is referred to Refs 2, 3 and 4. **To minimize pressure drop recourse must be** made to the use of bull nosed inlet section and tapered exit sections. Such designs can employ circular or parallel splitters depending on the duct geometry. Finally, it should be noted that they may on occasion be required to work under vacuum conditions and they need to be mechanically designed to withstand such conditions if the system design so requires.

8. Systems which are totally ducted, e.g. circulating systems in drier circuits, may not require the fitting of silencers since the duct-ing and any accompanying insulation which doubles for thermal and acoustic duties can produce the desired noise levels. For many systems only a suction silencer is required, e.g. the FD fan shown in Fig 1. Such choices can only be made after appraisal of the system in which the fan is employed and simple guidelines as to the choice of silencer and/or acoustic insulation **are best avoided as being a pathway to problems,** if not disaster.

9. Fan casings, if of sufficient surface area, can be large enough to emit significant noise power and it is common practice to use acoustic lagging to control such noise emission. A typical example of this can be seen in Fig 3: the before and after effects of the cladding are shown in Fig 4.

10. High temperature duties pose some difficulty in the choice of absorption material because many of the materials are resin-bonded and in some cases have a melting point close to operating temperature. To overcome this stainless steel wool has been employed. A curve showing its absorption coefficient as a function of frequency is shown in Fig 5 and from this it can be seen to have good acoustic properties.

CENTRIFUGAL PUMP NOISE

Prediction of pump noise

11. For most pump installations the actual noise levels generated by the pump are:

(i) less than those generated by the driver

(ii) sometimes swamped by noise from associated minimum flow bypass systems

(iii) conditioned by cavitation in the suction system

Clearly (iii) is a condition to be avoided whenever possible but (i) and (ii) are very much within the control of equipment and system designers. Fig 6 shows noise contours around a Boiler Feedwater Pump (375 kW 50 rps) and they clearly illustrate the noise generated by the drive motor is well in excess of that due to the pump casing. Fig 7 shows the noise generated by a Process Pump with a poorly designed minimum flow let down system and the end result achieved by a novel let down system and pipe lagging.

12. There are few references which provide means of predicting noise due to the pump alone and some over-predict the sound power levels by up to 20 dB. This may arise because the measurements used to produce the predictive method include noise from the extraneous sources mentioned above. From a few measurements taken on ICI installations it would appear that the following technique gives a realistic value for sound power emitted by a pump casing

$$L_w = K_2 + 70 + 10 \log_{10} kW \qquad (2)$$

where

K_2 = the octave band correction tabled below

kW = pump shaft power, kW

L_w = sound power in dB re 10^{-12} Watts

Octave Band Centre Frequency Hz	31.5	63	125	250	500	1000	2000	4000	8000
K_2	-6	-6	-6	-8	-8	-10	-15	-20	-20

Control of Pump Noise

13. The installation of silencers in centrifugal pump suction and delivery systems is not practical in that absorption materials will often fail to work in such systems and reactive silencers are impractically large for low cut off frequencies. In any event they are frequently unnecessary.

14. It is more profitable to design the driving motor and by-pass systems correctly **and use pipe lagging to minimize noise** generated by the rotor. For installations where experience or test bed results make it clear that noise from the casing is likely to be a problem then Acoustic Hoods provide a practical if unwieldy and non-preferred solution to the problem.

NOISE DUE TO LARGE CENTRIFUGAL COMPRESSOR INSTALLATIONS

Prediction of noise levels

15. There are three basic conditions which need to be considered

  (i)   normal operation

  (ii) operation at start up with abnormal flow and pressure ratios

  (iii) operation at normal flow conditions to vent or on total by-pass

Again each of these conditions is dependent on noise from the driving elements and system layout and design. It is worth noting that (ii) and (iii) can pose difficulties in the design of acoustic treatments.

16. For normal operation one is concerned only with internally generated noise and its emission from the machine casing and associated pipework and vessels. Experimental work has shown that casing sound power can be calculated approximately from the following method

$$L_w = K_3 + 10 \log_{10} A \qquad (3)$$

where

L_w = sound power in dB re 10^{-12} Watts

A = casing surface area, m^2

K_3 = octave band constant tabled below

Octave Band Centre Frequency Hz	31.5	63	125	250	500	1000	2000	4000	8000
K_3	85	88	91	84	81	85	84	69	62

There is no theoretical basis for this technique and it should not be applied to screw or reciprocating compressor installations.

17. Pipework noise can be evaluated by careful consideration of noise from the machine and that due to turbulent flow which is controlled by line diameter, flow velocity, density, and viscosity. For some installations the control of flow velocity is the most important factor and experience has shown that under-sizing of pipework leads to unnecessarily high noise levels. Full account needs to be taken of the effective use of any thermal lagging which may be fitted: by influencing the type and design of lagging an optimum performance for both thermal and acoustic behaviour can be obtained.

18. Internally generated noise is inevitably transmitted into suction and delivery pipework where it is substantially attenuated by pipe walls and lagging. Its importance should not be overlooked because the large pipe surface areas available for noise radiation can make it a major factor in overall noise emission. For centrifugal machines the noise power transmitted into the suction and delivery pipework is usually conditioned by the first and final stages respectively: the contribution from intermediate stages is usually contained within the machine by Mach No. effects. Various methods of calculation have been correlated with test results taken on Air Compressor inlets and it would appear that the equation given below, due to Piestrup and Wesler is some 15 dB too high.

For practical use a reduction of 15 dB is suggested.

$$L_w = 107 + 17.7 \log_{10} \frac{kW}{N} + 15 \log_{10} \frac{N}{6}$$

where

kW = power for stage considered

N = No. of vanes of the stage

None of the above considers pure tone noise and it is worth noting that pure tone noise is rarely observed with centrifugal machines. If it is observed then it is usually due to flow disturbances at the inlet or outlet of a machine. Typically pure tones can be generated by tight pipe bends close to the inlet of a circulator causing an asymmetric flow into the eye of an impeller or by tight cut off vane in an outlet diffuser. Fig 8 shows a noise spectrum from a circulator with a tight inlet bend configuration. The removal of flow asymmetry by straightening vanes is possible but not practical if the security of a machine is important. A more reliable technique is to ensure that there are at least 10 diameters of straight pipe after a bend: the cost and practicability of such an alteration on an existing plant should not be under-estimated for HP systems.

Control of compressor system noise

19. There are three techniques which can be applied

  (i)   provide adequate suction, delivery and vent/by-pass, silencers as required

  (ii) acoustically lag system pipework

  (iii) fit Acoustic Hoods to the compressor

These can be applied jointly or singly dependent upon the various acoustic and practical requirements: items (i) and (ii) require minimum change to compressor systems whereas (iii) does require detailed consideration at the outset of a Project, e.g. the building has to be sized to permit the hood to be placed in a "safe" place for construction and maintenance. Even so the design of the other silencers needs to be considered at an early stage: this is particularly important for suction and vent silencers on Air Compressors because of their relatively large size. Whilst the acoustic design is done by standard techniques well described in the literature it should be noted that the constraints of layout have a major influence on their design, i.e. compromise is inevitable.

20. **Suction and delivery silencers**

a   Suction silencers

    These used to handle full flow with minimal pressure drop and are usually of the absorption type. In an acoustic sense they are little different from those used on Fans. A design for an inline air intake compressor is shown in Fig 9 which also gives the calculated attenuation for the various octaves: it is worth noting that the need for some minor attenuation in the 63 and 125 Hz octaves produces excess attenuation in the 1000 and 2000 Hz octaves. Mechanical considerations are of prime importance and it is important that suction silencers have a basic structure strong enough to withstand negative pressures generated by blinding of the Air Intake Filter which incidentally can provide some of the attenuation and so limit the size of the intake silencer.

b   Delivery silencers

These are considerably smaller than suction silencers for most Process Machines and a sample design is shown in Fig 10. Their attenuation requirements are less than for open inlet air systems because of attenuation due to the pipe walls: permissible pressure drop although important is more readily achieved because the operating pressure is high. The main design constraint is that operation at reduced pressures during start up can lead to high flow velocities: the absorption material must therefore be contained in such a way as to prevent its egress into the process.

21. Acoustic lagging of pipework

The use of acoustic lagging for compressors is in principle much the same as that for fans (Section on Control of fan noise) and apart from the need to use thicker sections in some cases there is little more to add. Details of fixing vary with application and low temperature duties on refrigeration circuits require that vapour seals etc are properly employed.

22. Acoustic hoods

Acoustic hoods are now used extensively on portable air compressors and these are now part of the standard package offered by most manufacturers of this type of equipment. Process machines using similar but much larger hoods for the purpose of controlling noise have had more limited use because of cost, difficulties in retro fitting, safety reasons etc. Nevertheless, they are now coming into use where the need is strong enough.

23. Usually one is looking for up to 20 dBA reduction in noise level and examination of Fig 11 shows this is readily available. However, acoustic hoods do pose problems over a number of matters to do with machine operation, e.g. provision of a silenced ventilation system to avoid overheating and the build up of noxious fumes, the redesign of lubrication systems to minimize the effects of oil leakage into acoustic absorption and the possibility of "lagging fires". A complete account of this is given in Appendix 2 which presents a specification devised by Agricultural Division to ensure that such hoods are made to an adequate standard.

DRIVER NOISE

24. This is just as important in the control of noise from pump, fan and compressor installations as the driven machines themselves. There are usually only electric motor or steam turbine drives employed, often with the interposition of a gear drive, and occasionally for the smaller pieces of plant belt drives. It is not the purpose of this paper to examine these problems in the same depth but some briefer consideration is given below.

Electric Motors

25. By good electromagnetic design it is possible to avoid the pure tone noise which may arise due to rotor/stator slot effects and until recently one was then often left with noise from the

cooling fan. Thus we have returned back to Section on "Fan Noise" except that

(1)   many of these fans are bidirectional and hence their aerodynamics are of the most elementary form

(2)   inadequate matching between the required and actual fan capabilities has given temperature rises well within specification at the expense of noise

(3)   leakage in the air circuits can also lead to increased fan size and thus more noise: in one case the author deduced that at least 25% of the air in the fan was on simple recirculation due to "normal working clearances".

26. Over the past few years development work by manufacturers has improved on this gloomy situation by judicious changes in fan design and some extra acoustic treatment. To illustrate this the noise spectrum from a 2000 kW 25 rps motor is shown in Fig 12. For very large electric motors CACW systems are employed and these are of a low noise nature: this is also illustrated in Fig 12 where the spectrum for a 5000 kW 25 rps motor is shown.

Steam turbines

27. Much of the noise from these originates in control valves and lubrication consoles rather than from the insulated and shrouded casing. By proper insulation of the pipework this noise can be minimized. Subsidiary to this is the need to properly size and insulate steam inlet and exhaust systems.

Gearing and couplings

28. Gearing noise is characterised by the presence of impacting as teeth move into contact: for this reason gear noise is primarily a function of good design and fabrication. Experimental work within ICI suggests that the noise power output is also a function of size and in this respect compact epicyclic gearing offers some advantages amongst which is the flexible connection of the annulus to the casing. A noise spectrum taken on a test bed for an epicyclic box is shown in Fig 13. Despite this there is often a need to place gearing within an acoustic hood: the major consequence of this is that the associated oil cooler must be increased in size to make up for the loss of convective cooling.

29. A final word on couplings is useful because high peripheral speeds cause air turbulence and hence noise: for most situations the noise is contained within the coupling guard depending upon its degree of tightness.

COLLABORATION BETWEEN USER, DESIGN CONTRACTOR AND EQUIPMENT VENDOR

30. All of the foregoing sections have no meaning unless the User and Design Contractor combine to produce noise specifications which the equipment vendor clearly understands and is capable of achieving. In the past, too many specifications have been written and not worked to because of lack of will, expertise and, in some cases, wilful disregard. This situation is no longer acceptable and this conference, and the change in attitudes over recent years, is proof of this.

31. Specifications must be written with an eye to both what is desired and that which is achievable: if a particular situation calls for noise levels below that which best practice can achieve then

(1) the use of external acoustic treatments must be considered, e.g. Acoustic Hoods, Machinery Houses, or simple noise barriers

(2) alternative methods or equipment should be considered for the particular duty, e.g. use screw pumps rather than gear pumps for lubrication systems.

32. Experience gained in operation should be passed to both the Design Contractor and Equipment Vendors: the specification in Appendix 1 attempts to do this in a simple non-numerative way. Design Contractors and Vendors rarely ask about equipment they have supplied or even attempt to get data during commissioning and it is more usual for them to be concerned only with problems which may arise. It is the author's opinion that it is just as important to have knowledge of equipment which operates well as for the small percentage which give trouble.

33. Within both the User and Design Contractor organizations there is a need for the Noise Engineer to integrate himself into the Project Team and it is vital that he utilizes the Project and Functional Design Engineers to carry out much of the detailed contractual work. Experience has shown this can work providing the others properly call upon the Noise Engineer for assistance when problems arise: to this end it is essential that the Project Team are fully aware of the problems likely to arise in the specific areas they control. A major problem that has been revealed is failure of main vendors to pass on noise requirements to sub-contractors: the solution is purely a matter of administrative discipline and it certainly is not a technical problem.

CONCLUDING REMARKS

34. An attempt has been made to show the technical and some of the philosophical bases as to how Noise Control can be effected within a major project. It is not claimed that this is the only route which can be followed but it has been applied to a number of projects which have been successful over recent years.

REFERENCES

1. J B ERSKINE/J BRUNT. Prediction and control of noise in fan installations. Conference on Vibration and Noise in Pump, Fan and Compressor Installations, Institution of Mechanical Engineers, 1975.

2. L L BERANEK. Noise reduction. McGraw Hill.

3. I SHARLAND, Woods practical guide to noise control. Woods of Colchester.

4. E B MAGRAB. Environmental noise control. John Wiley & Sons.

APPENDIX 1

NOISE SPECIFICATION FOR PROJECT 2942

SCOPE

1. Noise limits for the Project are given in Sections headed "Noise levels within a Plant" and "Perimeter Noise" to provide the general framework for Acoustic Design. The Section on "Possible Problem Areas and Constraints" outlines possible problem areas and constraints and is included as a non-exclusive guide. For individual manufactured plant items, separate Noise Specifications are to be set up which satisfy the general requirements outlined in this document; these will be the subject of separate, and more detailed, agreements. The definitions for the various terms employed can be found in BS 661:1955 - Glossary of Acoustical Terms.

2. Agricultural Division are concerned in the light of previous experience, and impending changes in Noise Legislation, that the specified noise limits are not exceeded. It is expected that evidence will be supplied, in detail, by the contractor during bid discussion as to how they intend to achieve the stated objectives.

SPECIFICATION

Noise levels within a plant

3. In areas to which personnel require access in the normal course of their duties, the noise level from all sources shall not exceed that stated in Table 1 including abnormal conditions, e.g. venting and flaring, compressor start up on by-pass. These levels are given for each of the preferred octave bands and refer to continuous broad band noise. If the noise is expected to contain very narrow band or pure tone components, these levels are to be reduced by 10 dB in the octave which contains the pure tone. If the noise is impulsive in character the levels are to be reduced by 10 dB throughout, where the kind of instrumentation referred to in sub-section "Instrumentation" is employed.

Table 1

Limiting Sound Pressure Level in Decibels re 2 x 10⁻⁵ N/m²	Octave Band Mid Frequency	'On' Plant	Workshops	Control Room/Plant Offices	General Offices	Canteens	Private Offices
	63 Hz	102	87	78	75	71	67
	125 Hz	96	79	70	65	60	56
	250 Hz	91	73	63	58	54	49
	500 Hz	87	68	58	54	49	44
	1000 Hz	85	65	55	50	45	40
	2000 Hz	83	63	53	48	43	38
	4000 Hz	81	61	51	46	41	35
	8000 Hz	79	59	49	44	38	33
	dBA	90	70	60	55	50	45
Noise Rating		85	65	55	50	45	40

Attention is drawn to the Noise Rating Curves of Figure 14.

4. Table 1 sets out the limiting levels which must not be exceeded in locations on and around a plant. The 'On' Plant level is regarded as the level at which impairment of hearing will not occur from long term exposure whilst the remainder are nuisance levels. The octave band figures for the desired noise rating from Table 1 must be used to derive the specified noise levels to be specified to equipment suppliers in the specifications for manufactured plant items.

Perimeter noise

5. At specified points in the residential areas adjacent to the plants, the noise levels which are attributable to the plant working normally must be low enough to ensure that justifiable complaints will not arise and that the possibility of litigation is avoided. In this connection, it is probable that the draft legislation now being discussed will be in force before the plant is commissioned. In addition to this, it is essential that the occupants of Process Offices are not subject to noise nuisance. The limits in Table 2 have been formulated to satisfy these two requirements.

6. The sound pressure levels produced by the Plant at the specified point must not exceed that set out in Table 2 and the contractor must endeavour to achieve lower levels if

possible. This point is sufficiently close to the plant to permit deduction of the actual plant noise from changes in the existing noise level.

Table 2

Limiting Sound Pressure Level in Decibels re 2 x 10⁵ N/m²		Specified Points
		100M South of A
	31.5 Hz	74
	63 Hz	76
	125 Hz	72
	250 Hz	66
	500 Hz	63
	1000 Hz	60
	2000 Hz	57
	4000 Hz	54
	8000 Hz	49
	dBA	65

A: this is the geometric centre of the proposed plant site: 20' North of Flare Vent Stack shown on Drawing BM/A AP/S-41A.

7. During periods of plant upset the noise level due to venting can be permitted to rise to 70 dBA at the exterior of Process Offices without inducting vigorous complaints from its occupants. The vent system noise 100M south of A should not exceed the limits set in Table 3 if the above mentioned 70 dBA is not to be exceeded.

Table 3

	Specified Point
	100M South of A
31.5 Hz	82
63 Hz	84
125 Hz	80
250 Hz	74
500 Hz	71
1000 Hz	68
2000 Hz	65
4000 Hz	62
8000 Hz	57
dBA	73

Limiting Sound Pressure Level in Decibels re 2×10^{-5} N/m^2

NOTES

Treatment

8. Where noise levels are expected to exceed those given in Table 1 or Table 2 the Contractor shall discuss with the Company the proposed methods of treatment to meet the stated limits: in this connection it should be noted that ear protectors for plant operators are acceptable only for cases of occasional short term exposure. To establish likely problem areas clearly the contractor shall prepare a note defining the likely noise sources, their octave band sound pressure levels and the resultant octave band sound power levels. Against each source a description of the means of reducing noise shall be outlined for the purposes of subsequent discussion: this must be accompanied by a statement of the expected cost of acoustic treatment.

Instrumentation

9. The octave band levels are to be measured with a sound level meter which complies with BS 4197:1967 and a filter set which complies with BS 2475:1964. Linear or weighting characteristic C is to be used, the response set to 'slow' and the maximum reading at each measurement point is to be taken. If the filter is adjustable, it must be set up in accordance with the makers' instructions to provide a linear characteristic. The weighted sound pressure level is to be measured with the weighting characteristic A.

Pure tone

10. A pure tone is present in the noise when a component in a given octave band is within 10 dB of the total level in the chosen octave band. Such a pure tone would be detected normally by ear but special instrumentation is necessary if a quantitative analysis of the tone is required.

POSSIBLE PROBLEM AREAS AND CONSTRAINTS

Previous experience with noise on ammonia and other similar plants

11. Experience has shown that the following plant items are major noise sources; they are given as a non-exclusive guide to expedite cooperation and assist in the acoustic design of the plant:

(1) Fin Fan Coolers.

(2) Naturally aspirated furnaces and their burners: this is particularly important when the fuel gas can have a low MW due to H_2 being present.

(3) Steam and gas pipework subject to high velocities: Annexe 1 gives a table of preferred velocities.

(4) Pressure let down systems and venting systems.

(5) TEFC and CACA motor cooling fans.

(6) Large fans in cooling towers.

(7) Control valves with pressure drops exceeding 10% of system pressure: this is particularly important with steam/gas mixing systems.

(8) Heat exchangers whose acoustic resonances are excited by flow noise set up by vortices.

(9) Gear pumps in lubricating oil systems.

(10) Lobe and screw compressors.

(11) Large centrifugal compressors, and prime movers.

(12) Centrifugal pumps with large minimum flow kick-back systems which produce 2 phase flow in low pressure pipework whose diameter equals that of the high pressure section.

(13) Gearboxes are frequent source of high noise levels and in the past recourse has been made to ground gear teeth or easily fitted acoustic hoods to overcome this effect.

12. The major constraints for this Project are likely to be the 'on plant' condition which defines exposure of plant operators to noise and the proximity of the plant to Process Offices. With respect to the latter constraint the vent systems will require particular attention.

ANNEXE 1

SOME PRACTICAL GUIDANCE ON PIPELINE VELOCITIES

13. The complexity of the subject precludes any absolute simple guide as to noise levels. In particular, the effects of density, viscosity and pipe diameter prevent a simple criteria based on economic velocities. However, the

following velocities are suggested as starting points when designing against noise: detailed checks can then be carried out using methods available in ICI.

14. Gaseous systems

a   Pipe diameters up to 75 mm
Velocity should not exceed 60 m/sec

b   Pipe diameter between 75 mm and 300 mm
Velocities should not exceed 30 m/sec

c   Pipe diameters above 300 mm
Velocities should not exceed 25 m/sec

15. Liquid systems

The commonly used velocities of up to 2 m/sec pose no problem on normal plants but for office blocks velocities should not be more than 1 m/sec.

APPENDIX 2

SPECIFICATION OF AN ACOUSTIC HOOD FOR NOISE ABATEMENT FOR PROCESS AIR COMPRESSOR

ACOUSTIC RATING

1. The hood shall complement the noise reduction measures incorporated in the machine installation and foundation design.

2. These measures are:

a   provision of monolithic foundation

b   complete enclosure of foundation sides using:

- solid fill between pillars: this may be in the form of brickwork or suitably damped steel panels

- acoustic louvres where natural ventilation is to be applied

- rubber seals at points where piping emerges from the in fill seal panels

c   provision of silencers in gas lines to and from the machine

d   exclusion of all ancillary equipment requiring process or routine maintenance attention from the enclosed foundation.

3. The installation shall be designed to achieve a noise level of less than ISO Noise Rating NR* under free-field conditions measured 1 m from the hood, over the specified range of duty. This level has been chosen to ensure that the DoE Code of Practice is adequately met in the final plant configuration.

4. Where no field data exists it will be assumed that pure tones exist in octave bands with major rotational, tooth contact, and blade passage frequencies. This shall be compensated for by a 5 dB reduction in the allowable levels.

GENERAL ARRANGEMENT

5. The hood shall be reasonably close-fitting to the machine casings and be designed for

*to be defined after full consideration of plant requirements

direct lift and transport by crate to a temporary storage area during periods of machine maintenance and inspection.

6. The machine operation shall not rely on the fitting of the hood. During commissioning after initial erection, or after maintenance, the machine will be run without the hood in place for purposes of inspection and test.

7. It must be possible to lift and replace the hood without disturbing piping or instrument connections.

8. Sectioned hoods shall be bolted together for manufacturing convenience and transport purposes to limit the size of parts to a maximum of 2.9 m wide 3.5 m high 12 m long.

9. Divided hoods may be employed on long machine trains, lifting in sequence for access to the appropriate machine casing.

LIFTING

10. The hood, or each section of a divided hood, shall be provided with lifting tackle suitably disposed about the centre of gravity to ensure a level lift.

11. The position of the centre of gravity and the weight shall be clearly marked on each hood section.

VENTILATION

12. Either natural or fan ventilation may be used to ensure:

a   at least 60 changes/hour of the air enclosed within the hood.

b   an air flow sufficient to limit the air temperature at any point within the hood to less than 70°C, assuming the compressor house ambient air temperature is 32°C and the worst condition for leakage from the machine seals.

13. For fan ventilation a forced draught system is preferred. The hood aperture sizing for the exit air shall take into account the leakage gas flow from the labyrinth seals on the main compressor at the maximum clearance permitted for wear.

14. Where induced draught fans are necessary they shall be of the radial flow type with pressure/volume characteristics sufficiently flat to cater for this worn labyrinth condition whilst maintaining a sub-atmospheric pressure within the hood.

15. Both inlet and exit ventilation ducts shall be acoustically lined for silencing. The exit air shall emerge through slots or ducts at a level not less than 2.5 m above the machine operating platform. The silencers will have an acoustic performance at least equal to that of the hood.

16. Acoustic louvres used where natural ventilation is to be applied shall have an acoustic performance at least equal to that of the hood.

ACCESS

17. Access doors are not required. Windows shall not be fitted. Handholes for special access or ports for viewing may be provided only with the express approval of the Purchaser.

18. When provided such handholes shall be 250 mm diameter, closed by a blank flange sealed with a neoprene rubber gasket or 'O'-ring.

CONSTRUCTION

19. The shell shall be fabricated by welding from mild steel sheet not less than 2 mm thick suitably stiffened to eliminate drumming. Assemblies using self-tapping screws shall not be used.

20. To permit personnel access the shell frame and top plating shall hold a 150 kg load at any point.

21. If special access is required to lifting points then catladders shall be provided.

22. The lower edge of the hood shall be further stiffened or made of thicker steel sheet to resist kicking and shall be provided with a continuous, shaped, neoprene rubber seal to mate against a smooth level surround to the machine baseplate.

23. Where the hood goes over a coupling to another machine or driver; the coupling guard shall incorporate the fixed shell section necessary to permit a continuation of the hood lower edge seal to ensure acoustic sealing.

ACOUSTIC MEDIA

24. The acoustic media shall be mineral wool protected by fibreglass scrim cloth and by a perforated metal sheet. The media shall not be less than 50 mm thick generally and not less than 20 mm thick over inward protrusions from the shell and shall finish at least 150 mm above the lower edge of the hood.

25. The perforated sheet shall have at least 20% open area and shall not be less than 0.5 mm thickness in either 99.5% aluminium or galvanised mild steel.

SPECIAL REQUIREMENTS FOR THE MACHINES AND THEIR OIL CONSOLES CONSEQUENT UPON USE OF NOISE HOOD

Oil systems

26. All oil reservoirs and oil consoles shall be located outside the hood and immediate machine foundation. Shaft driven oil pumps and baseplate oil reservoirs are not permitted.

27. Sufficient oil must be supplied to bearings to ensure that the bearing housing exit oil temperature does not exceed 70°C.

28. Oil lines within the hood shall be fabricated by welding, leaving the fewest joints consistent with dismantling for machine maintenance. Joints shall be flanged except for sizes less than 20 mm where screwed connections are permitted provided they incorporate 'O'-ring or gasket seals.

29. Oil mist escape from bearing housings shaft seals shall be prevented. Approved methods of complying with this requirement are:

- provision of purge air to the centre ring of each bearing housing labyrinth shaft oil seal. Such purge air can be bled from the main air compressor provided the air temperature does not exceed 65°C under any condition of operation. Alternatively the purge air can be obtained from an external fan through a filter rated at more than 85% efficiency as measured by the AFI dust spot test.

- provision of an extractor fan from the oil reservoir, with the fan exhaust led through a catchpot with wire mesh demister before discharge to atmosphere. Pressurizing the hood for this purpose is not acceptable. Breathers shall be eliminated, or piped direct to the oil reservoir.

Instrumentation and control system actuators

30. Valve, inlet guide vane or stator blade actuators mounted within the hood may be pneumatic or electric servomotors. Hydraulic systems using oil are not permitted.

31. Axial/centrifugal compressors without intercoolers or aftercoolers must be fitted with surge pulse detection trips to avoid the high gas temperatures arising from surge.

32. All temperature and vibration sensors mounted within the hood shall be triplicated. All oil, gas, steam or water pressure or flow sensing shall be located outside the hood.

33. The air temperature at the exit from the hood shall be measured and indicated.

Insulation

34. Insulation shall be applied to all machine and pipe surfaces within the hood where the surface temperature can exceed 150°C. Such insulation shall be of the crumpled aluminium foil type, arranged for self-draining. Fibre type insulation shall not be used.

Fig. 1:

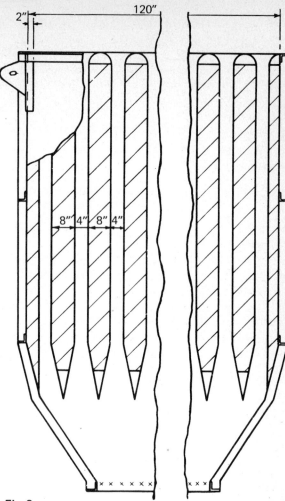

Fig. 2:

Fig. 3:

32

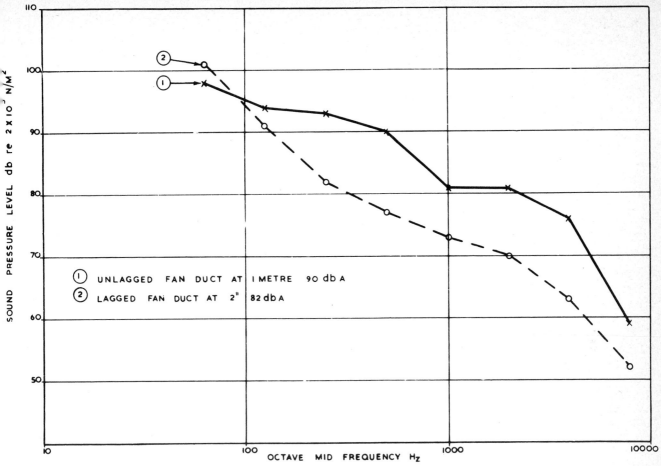

① UNLAGGED FAN DUCT AT 1 METRE 90 db A
② LAGGED FAN DUCT AT 2" 82 db A

Fig. 4:

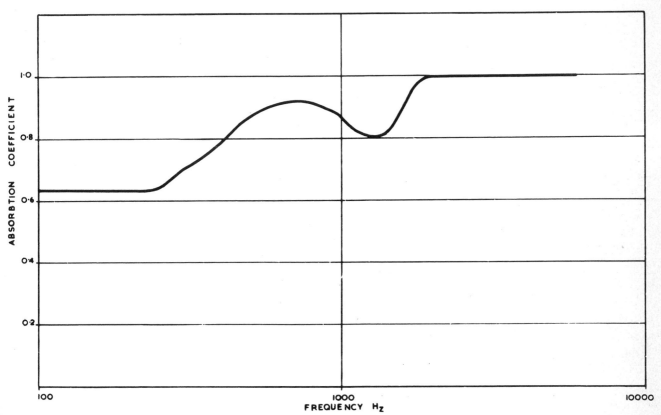

Fig. 5:

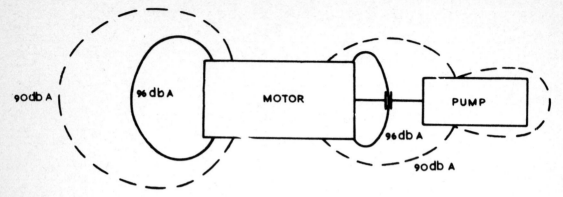

Fig. 6:

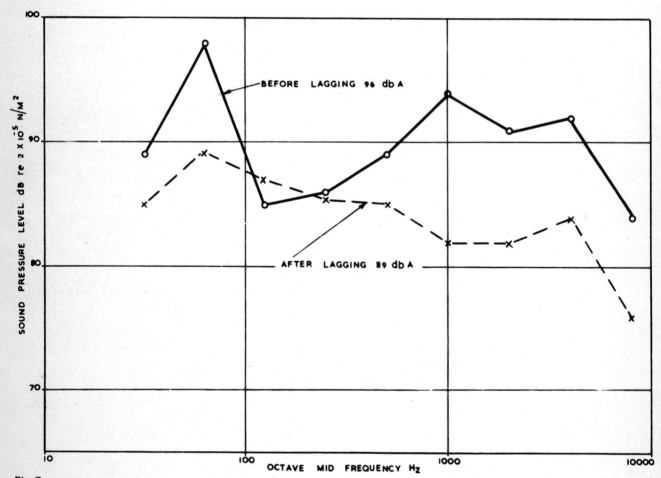

Fig. 7:

34

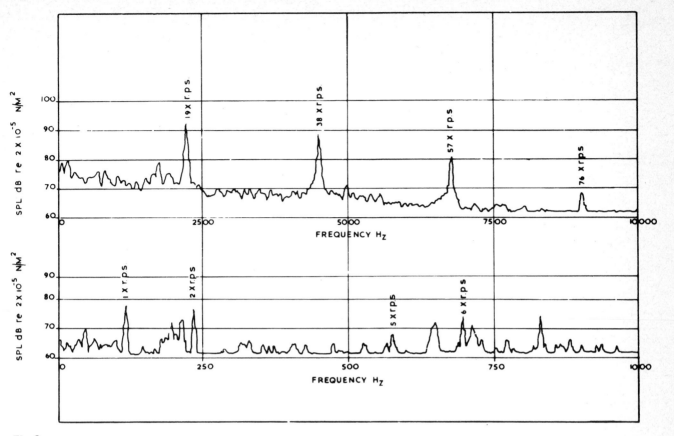

Fig. 8:

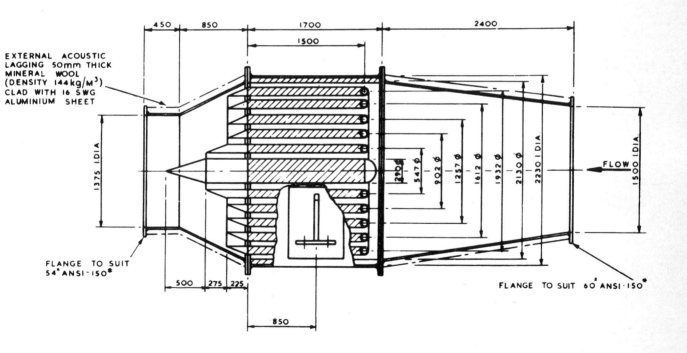

EXTERNAL ACOUSTIC LAGGING 50mm THICK MINERAL WOOL (DENSITY 144 kg/M³) CLAD WITH 16 SWG ALUMINIUM SHEET

FLANGE TO SUIT 54" ANSI·150#

FLANGE TO SUIT 60" ANSI·150#

FLOW

CALCULATED PERFORMANCE

OCTAVE BAND MID FREQUENCY Hz	63	125	250	500	1000	2000	4000	8000
INSERTION LOSS dB	10	11	27	42	54	56	51	33

Fig. 9:

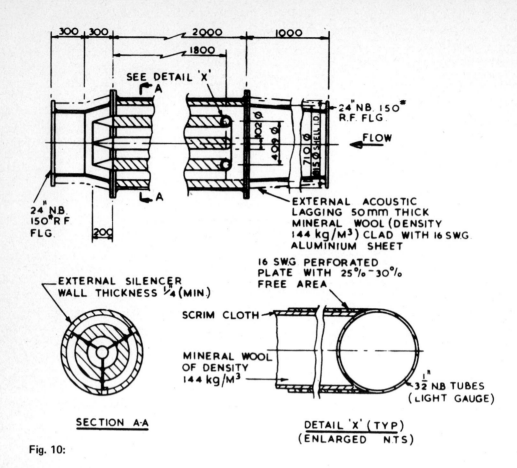

Fig. 10:

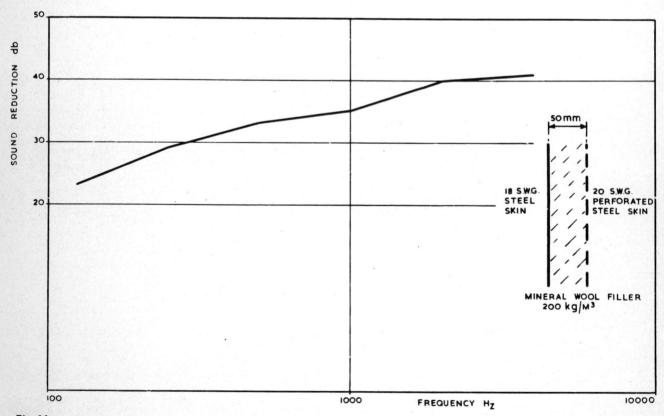

Fig. 11:

36

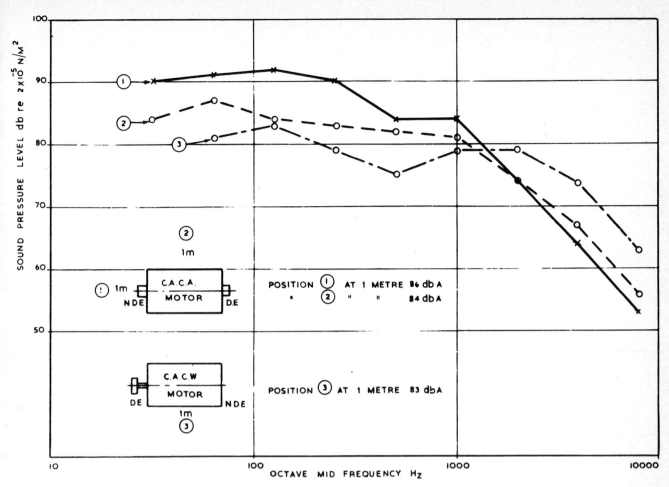

Fig. 12:

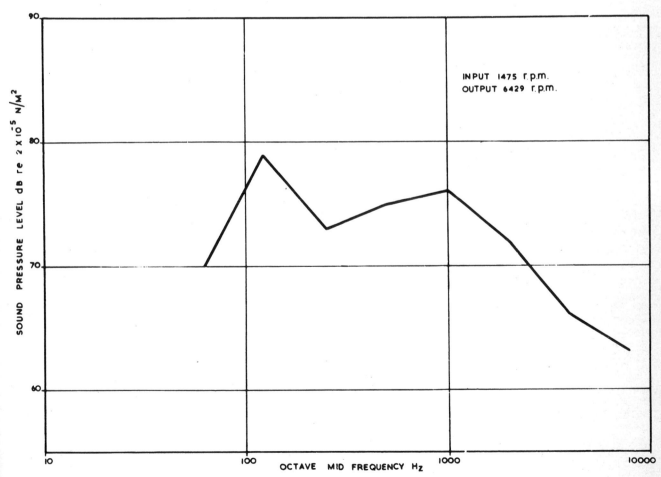

Fig. 13:

37

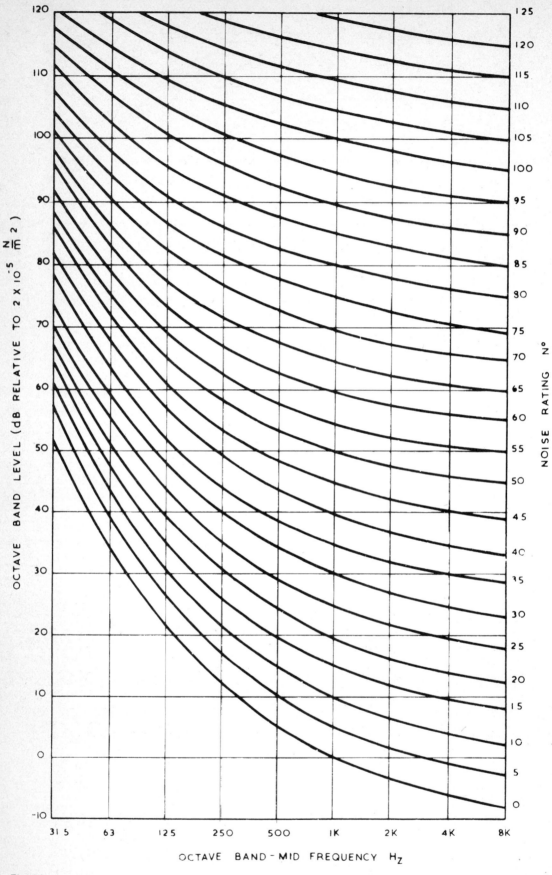

Fig. 14:

C250/77

RECENT WORK BY CONCAWE ON THE MEASUREMENT OF NOISE EMISSION FROM LARGE SOURCES

K.J. MARSH, BSc, MIstnP
British Petroleum Co.

SYNOPSIS   The standard procedures currently being considered by the ISO for determining the sound-power levels of machines are not suitable for large sources such as furnaces, banks of cooling fans, and large compressors. For some years CONCAWE has recognized that there are intrinsic errors involved in making measurements in the near field of such sources and has made a study of these errors. The results of this study have been used to develop new measuring procedures for determining the sound-power levels of furnaces and cooling fans: similar procedures could also be developed for large compressors and gas turbine sets. The paper discusses the problems involved in making measurements and indicates how they are resolved in the test procedures.

NOTATION

A	Acoustic efficiency of vibrating surface (radiation ratio)
A_f	Arctan (R_f/d)
A_h	Arctan (R_h/d)
d	Measuring distance
L_p	Sound-pressure level
L_v	Vibratory-velocity level $(L_v = 20 \lg v/v_o)$
L_w	Sound-power level
R	Radius of circular source
R_f	Radius of fan
R_h	Radius of fan hub
S	Area of measuring surface
s_o	Reference area
v	Root-mean-square vibratory velocity
v_o	Reference velocity level

INTRODUCTION

1.   In 1970 the Oil Companies Materials Association (OCMA) published its Specification NWG-1, which contained procedures for measuring the noise emission from small and large sources (Ref. 1). These are essentially practical procedures which can be applied in the unfavourable surroundings of factories or petroleum plants, where most measurements on machines and equipment have to be made. Various test procedures have been proposed in recent years for measuring the noise from small sources (usually as sound-power level) and these include procedures from the International Standards Organisation, national standards such as German DIN standards, and specific procedures for special types of equipment. Unfortunately, many of these tend to be complicated – particularly in the number and choice of measurement positions – and although designated as "engineering" methods are not practical for most engineering requirements; they are also limited in the size of source to which they can be applied.

Very little attention has been paid to the problem of measuring noise from large sources – despite the lead set by OCMA NWG-1 – and yet these are often the most significant contributors to both work-area noise and neighbourhood noise. Such sources include furnaces, banks of air-cooler fans, and large compressors or gas turbine sets; often these can only be tested at full capacity when the entire plant is operating. This means that there is considerable background noise and significant measurements can only be made close to the source, effectively within the acoustic near field.

It was recognized during the preparation of OCMA NWG-1 that such measurements in the near field of large sources would involve errors in the determination of sound-power level, but they were deliberately ignored in the interests of developing a practical procedure in a reasonable time. In many circumstances the errors were not apparent because they could not be separated from other measuring errors due to ground absorption and the screening effect of other equipment or buildings. But, in recent years, detailed studies on air-cooler fan test rigs have shown that there is a consistent discrepancy between sound-power levels derived from near-field measurements and those derived from far-field measurements; the latter are usually found to be 3 – 5 dB lower. This is large enough to affect the selection of large equipment for installations where there are severe environmental restrictions.

It was clear that more information on this problem was needed and a detailed study was commissioned by CONCAWE, the international environmental study organisation of the European oil companies. This paper describes briefly the results of the CONCAWE study and its application to the development of new test procedures for large sources.

2. Although the acoustics of the near field have been described in the specialist literature, there are few summaries sutiable for the non specialist, so CONCAWE engaged a noise consultant, Müller-BBM, München, to prepare a literature survey and to comment on the problems relating to the practical determination of sound-power levels. This has now been issued as a CONCAWE report (Ref. 2).

The report distinguishes two features of the near field. The first of these is the "acoustic" near field, where the sound pressure and particle velocity are not in phase; the second is the "geometric" near field where the large dimension of the source in relation to the measuring distance causes sound waves to travel in directions which are not normal to the measuring surface. Both of these cause an increase in the apparent sound-power level derived from measurements of sound-pressure level in the near field.

The correction for the acoustic near-field error is difficult to estimate in general; it depends, for example, on whether the source is dipole or quadrupole and on the way the sources are distributed across the surface of a machine. However, in most practical cases the error is less than 1 dB for the 125Hz and higher bands, provided that the measuring distance is 1 m or greater; for the 63 Hz band the measuring distance has to be at least 2 m.

Many test methods require a minimum measuring distance of 1 m but it may be noted that in the draft ISO standard DIS 3744.2 the minimum distance is 0.25 m, and 1 m is merely the preferred distance; measurements made at such a small distance must therefore be checked for near-field errors by a further set of measurements at a larger distance. This doubles the number of measurements – which is already large.

It is easier to estimate the geometric near field error for a number of typical situations because it depends mainly on the angle subtended at the microphone by the radiating surface. If, for example, the source is a diffusely radiating wall or opening it is usually assumed that the sound-power level, Lw, of such a source is given by

$$Lw = Lp + lg(S/so)$$

where Lp is the sound-pressure level measured near the surface, or in the opening, and S is the source area.

But, in fact, this overestimates the sound-power level because it assumes that all the sound energy is moving normally through the measuring surface. If the surface is treated as an array of incoherent sources and consideration is given to the sound energy arriving at a wide range of angles to the normal, an integration can be made for the effect of the extended surface area. A cosine factor must be used for the directivity of the sources because if they are taken as omnidirectional (as in Maekawa's analysis) the sound-pressure level becomes infinite at zero distance and this does not occur in practice. For a circular opening, the amount of the overestimate is found to have the following values:

Error(dB):	3	2.6	2.2	1.4	1.0	0.7
Half angle(deg)	90	84	79	63	55	45
d/R	0	0.1	0.2	0.5	0.7	1.0

where the half angle is the angle subtended at the microphone by the radius, R, and d is the distance of the microphone from the surface.

This shows that, when the microphone is placed in an opening from a noisy room or near to a large radiating wall, the error in the estimate of sound-power level is 3 dB. As the microphone is moved away from the source the error decreases, but it could still be significant for measurements made at 1 m from the surface of large machines using the various published small-source procedures. It is an aspect which needs consideration in these test methods.

A similar calculation shows that when measurements are made close to a line source, such as a radiating pipe, there is an overestimate of 2 dB if the sound energy is merely assumed to be radiating normally through a cylinder. The error changes as the microphone is moved away from the source and Maekawa has given the equation and graph of the variation (Ref. 3).

These errors can be significantly reduced by using directional microphones since, in principle, these measure only the energy moving normally through the measuring surface. In practice, the directional properties vary with frequency, being considerably less pronounced at low frequencies. Directional microphones also have the disadvantage that they are not commonly available and are not compatible with some precision sound-level meters; they are also more difficult to calibrate. All in all, they are not convenient for use by the non specialist.

THE USE OF VIBRATION MEASUREMENTS FOR ESTIMATING NOISE EMISSION

3. Some precision sound-level meters are supplied with vibration transducers so that the instrument can be used to measure vibrations in the acoustic frequencies. Unfortunately, the operating instructions do not always give a clear guidance on how to convert the vibration measurements to a quantitive prediction of the sound radiation from a surface; also the literature references, for example Cremer and Heckl, are not readily accessible (Ref.4). As a result, many engineers merely use their vibration attachments for a qualitative comparison of noise sources. In fact, there are many practical situations where vibration measurements can give a fairly reliable measure of the sound-power level of a source; it is particularly valuable where high background noise makes it impossible to take acoustic measurements.

The most convenient relationship is between the vibratory-velocity level, Lv, and sound-power level, Lw, which can be represented by the equation,

$$Lw = Lv + 10 \ lg \ S/so + 10 \ lg \ A$$

where S is the area of the vibrating surface.

In this equation the vibratory-velocity level, Lv, is defined in relation to the reference level, $v = 5 \times 10^{-8}$ m/s, and it should be noted that this is not the reference velocity recommended by the ISO but the one which gives a simpler relationship between Lw and Lv.

The difficulty in using these equations is the choice of a value for the radiation ratio A. However there are many practical situations where A is close to unity and its logarithm is therefore zero. This occurs, for example, when the vibrations in a flat plate are excited by air-borne sound waves, so the technique can be very valuable in measurements on furnace walls and on ducting; it can also be used for the radiation from pipes above the ring frequency.

This technique does not find favour in published standards because usually they are aiming at a level of precision which is not attainable in many practical circumstances. But for most engineering purposes, a lower level of precision is frequently acceptable - often, it represents the only realistic approach - and the vibration technique can then be useful. There are, of course, limitations in the application of the technique but, as its advantages are not widely appreciated, CONCAWE has published a report which gives a practical guidance to non-specialists on how to exploit the full potential of their sound-level meters with vibration attachments (Ref. 5).

NOISE TEST PROCEDURES

4.

Furnace Noise

4.1 Although furnaces do not come within the scope of this conference it is of interest to see how the concepts described above are applied to the measurement of their noise emission because the same principles are relevant to other large sources such as compressors. An example of this is the test method recently published by CONCAWE for furnaces used in the petroleum and petrochemical industries (Ref. 6).

The first aspect of this test method is that the furnace is treated as a number of noise sources whose sound-power levels are calculated separately; they include, for example, the furnace walls, forced-draught fans and their ducting, and the burner area (for floor-fired furnaces). Most of these have to be assessed from near-field measurements because the furnace can only be run at full capacity when the rest of the plant is operating, so background noise is always a problem. This means measuring near large radiating surfaces, such as the walls or the imaginary peripheral surfaces underneath the furnaces, and the geometrical near-field correction invariably has its maximum value of 3dB.

However, it is rarely practical to measure the noise radiated by the walls, or even some of the ducting, due to background noise, although the large areas involved could mean that they have a significant sound-power level. Measurement of their vibratory-velocity levels therefore represents the only possibility of determining their noise emission.

Another aspect of this test method is that it does not prescribe a rigid set of measurement positions; some discretion is left to the engineer and the method recommends a preliminary survey of the sound field around the furnace, using the sound-level meter set to dB(A) and the subjective impression of the ears. Some positions have to be eliminated because of neighbouring noise sources, and the emission from some areas has to be estimated by reference to other similar surfaces. This is the kind of approach which is essential to practical engineering test methods.

Fan Noise

4.2 The most common use of fans in the oil industry is for air-cooled heat exchangers and the fans are usually axial, with a diameter of of 3 - 4 m; a typical refinery might have 60 - 70 such fans and some installations may have 200 or more. Because of their number, fans can be a significant contributor to the noise radiated to the neighbourhood and this means that the noise measurement is a crucial issue. Unfortunately, there is no agreed method of measuring air-cooler fan noise.

The standard methods for ventilation fans are not satisfactory because they are based on noise measurements in ducts, whereas air-cooler fans radiate noise into a free field on one side and through a bundle of cooler tubes on the other side. A general method for measuring the noise was proposed in the OCMA specification NWG-1 as part of the large-source procedure and this is regarded by most oil companies as the most satisfactory test method. It is, however, subject to the near-field errors discussed above and a clear definition is required of the corrections which are necessary and of the specific application of the large-source method to fans, so CONCAWE is now preparing a standard test method.

This method is still in draft but it will require separate determinations of the sound-power level on the bundle side and on the fan side from near-field measurements. A special problem arises on the fan side of induced-draught fans because of the air flow, and high noise levels are measured if the microphone is too close to the fan. It is therefore necessary to put the microphone 1 - 3 m from the fan on the fan axis. There is then some question about the appropriate area to be taken for the measuring surface.

In the large-source method this would normally be the area of the radiating surface but as the measuring distance increases this is no longer valid; and if the fan has a large hub this would not be a part of the radiating surface. An investigation of this problem, based on the principles described in Ref. 2, suggests that the following equation should be used to determine the sound-power level of a fan radiating in free field with no significant directivity:

$$Lw = Lp + 10 \lg S/so - 10 \lg \left[2 \frac{(\cos Ah - \cos Af)}{(\sin^2 Af - \sin^2 Ah)} \right]$$

where $S = (Rh^2 + d^2)(Rf^2 + d^2)/d^2$

In this equation the third term is an expression of the geometric near-field error, which will range from 1 – 3 dB for most fans, and the second term contains the surface area; it is a hypothetical surface which has no physical significance. The test method will contain a graphical solution of the third term to avoid tedious computation for each fan test.

This equation represents a new approach to measuring the noise of fans radiating in a free field and it remains to be confirmed by detailed tests, but preliminary tests made so far indicate that it has a practical validity.

GENERAL

5.   The test methods described above illustrate some of the requirements for measuring the noise emission from large sources and CONCAWE recognises that there is a need for similar test methods for compressors and pumps. There is a problem common to both of these machines in that the noise they generate is radiated by the machine and the associated piping; frequently, the noise from the piping is more significant than that from the machine casing. A test method must distinguish between the two, probably by treating them as separate sources as in the CONCAWE furnace test method.

To measure the casing noise it is clearly necessary to separate it from the noise of the driver and the piping – particularly if there is a valve in the circuit. The piping noise however will depend on the material and dimensions of the pipe, also on whether it is on the discharge or suction side of the machine. This means that the vendor will have to devise a system for predicting the piping noise which includes the relevant parameters, so that the test measurements can be extrapolated to the practical situation. Already, valve manufacturers use this principle in their methods for predicting valve noise and until it is adopted in test procedures for pumps and compressors the test results will have little significance.

Predicting piping noise is now one of the biggest problems in the design of petrochemical plants, whether it is generated by pumps, compressors or valves, and it is now recognized that this is one of the major sources which controls the lowest noise levels that can be attained in such plants. It is a problem which could be partially solved by developing suitable test methods for pumps and compressors.

REFERENCES

1.   Oil Companies Materials Association. Procedural specification for limitation of noise in plant and equipment for use in the petroleum industry. Revision 1, October 1972.

2.   CONCAWE. Determination of sound-power levels of industrial equipment, particularly related to oil industry plant. Report No. 2/76. January 1976.

3.   Z. Maekawa. Noise reduction by distance from sources of various shapes. Appl.Acoustics (3) 1970, 225-238.

4.   Cremer, L. Heckl, M. Körpershall – Physikalische Grundlagen und technische Anwendung. Springer Verlag 1967.

5.   CONCAWE. Measurement of vibrations complementary to sound measurements (in preparation).

6.   CONCAWE. Test method for the measurement of noise emitted by furnaces for use in the petroleum and petrochemical industries. Report No. 3/77, January 1977.

C25I/77

CURRENT INDUSTRIAL PUMP AND FAN FLUID-BORNE NOISE LEVEL PREDICTION

W.M. DEEPROSE, MSc, ARCST and A.N. BOLTON, BSc
National Engineering Laboratory, East Kilbride

The MS of this paper was received at the Institution on 20th June 77 and accepted for publication on 8th August 1977

SYNOPSIS Measurements of fluid-borne noise levels have been taken from a wide range of industrial fans and pumps. Correlations have been derived relating the overall levels and frequency spectra to basic fan and pump parameters, and these allow reasonable estimates of fluid-borne noise to be made at an early stage of machine or system design. Some noise reduction techniques are discussed, which are applicable to both fan and pump systems, and experimental results are presented showing the varying degrees of success which the practical application of the techniques achieved.

INTRODUCTION

1.   The widespread use of fans and pumps and the demand for a controlled environment has generated a requirement for simple methods of noise level prediction useful to system engineers. This is particularly true of all types and sizes of fans in the ventilation and process industries where noise is introduced directly into sensitive domestic, commercial, residential and industrial environments. Often, noise control of fans can be achieved by non-acoustical methods through attention to the aerodynamics of the system in which the fan operates. Although pumps are as widely used, the requirement for fluid-borne noise level prediction is less obvious. However, there are certain specific areas where information on fluid-borne noise levels is important and reduction of these levels desirable.

2.   For both pumps and fans accurate and reliable measurement of noise level is essential in providing the basic data used to establish empirical prediction formulae. This paper therefore starts with a very brief statement of noise measurement procedures currently used for fans and pumps. Thereafter, a wide range of pump and fan data is presented in simple forms which provide a quick means of noise level prediction. Lastly, certain simple guidelines for minimizing noise levels from pump and fan systems are presented to enable system engineers to appreciate the substantial effects that simple engineering system choices may have on noise.

NOTATION

B	Blade number	–
c	Speed of sound	m/s
D	Impeller tip diameter	m
d	Impeller tip chord	m
f	Frequency	Hz
Δf	Frequency bandwidth	Hz
gH	Head rise	J/kg
L_W	Sound power level	dB re 10^{-12} W
M	Mach number	–
M_{20}	Mach number at 20°C	–
N	Rotational speed	rad/s
N_s	Specific speed	–
P	Pressure	N/m²
Q	Flowrate	m³/s
r	Rotor harmonic	–
t	Outlet tip width of centrifugal impeller	m
V	Number of stator vanes	–
$V_{m,2}$	Outlet meridional velocity	m/s
V_t	Tip speed	m/s
α	Incidence	degrees
ω	Rotational frequency	rad/s

PUMP NOISE PREDICTION

Pump noise measurement

3.   Measurement of pump fluid-borne noise levels is still a relatively new procedure and as yet no standard method has been determined. The data reported herein are based on measurements taken with flush-mounted hydrophones positioned often a few centimetres from the pump suction and discharge flanges. The pump circuits used for these measurements were those used for conventional performance testing and ranged in size from 10–50 cm diameter and from 10–100 metres in length. Therefore, although the acoustic loading on the pump and the effects of standing waves, bends, transitions and valves on the measurement uncertainty are acknowledged, they are not taken into account in the data presented here.

4.   In forming an estimate of the fluid-borne noise and the associated spectra three steps only are necessary:

a   estimating overall sound pressure level (OASPL) at the nominal best efficiency point,

b   correcting this level for flowrate, and

c   using simple geometric parameters to estimate the spectrum levels.

Predicting OASPL

5.   Measurements have been taken in nine pumps covering a specific speed range of 4:1 where specific speed is defined as

$$N_s = \frac{\omega \sqrt{Q}}{(gH)^{\frac{3}{4}}}. \qquad (1)$$

6.   The range presented covers the circulating and boiler feeder pump types and includes data taken on both the inlet and outlet side of the pump. Fig. 1 shows the overall sound pressure level measured at the best efficiency point for frequencies above 25 Hz plotted against the outlet tip speed in conditions where the pump performance was not affected by cavitation. The data, excepting the low-speed data from pump A, correlate to within ±4 dB for all other pumps. Pump A was a multi-stage boiler feed pump and, as the earliest pump to be tested, had the hydrophone located somewhat differently from later procedures. In this test the hydrophone was placed in a pitot traverser and located well into the flow stream. Other attempts to correlate these data based on sound power and on different combinations of pump size and speed produced less satisfactory correlations than that of Fig. 1.

Effect of flowrate other than at best efficiency

7.   Pumps are often required to operate at flows other than that of peak efficiency and it is usually found that the noise level varies with duty. Typical variations of OASPL with percentage flowrate are shown in Fig. 2 where a pattern of minimum OASPL around best efficiency point emerges. These data taken with that of Fig. 1 allow the OASPL to be estimated over a range of flow conditions from about 40-140 per cent flow.

Spectrum levels

8.   Attaching a spectrum to these OASPL can be achieved by plotting spectrum shape, that is sound pressure relative to OASPL, for each pump at each duty condition provided a suitable shaping parameter can be identified. An effective correlation has been achieved by producing a non-dimensional frequency, F, using outlet tip width t, and the outlet meridional velocity $V_{m,2}$, and dividing this dimensionless parameter by the pump specific speed, N_s. The reduced frequency may be written

$$F = \frac{ft/N_s}{V_{m,2}} \qquad (2)$$

The relative spectrum levels corrected by the parameter $(t/N_s)/V_{m,2}$ have been plotted against F to obtain Fig. 3. The data for this has not been shown for convenience since there were some 400 data points from nine pumps. The frequency range covered extends from 16 Hz-100 kHz. The bandwidth of ±6 dB contained about 90 per cent of the data plotted whilst threequarters of the remainder lay slightly below. It must be noted

that Fig. 3 refers to pump conditions which are truly non-cavitating as defined by a high frequency noise test to identify inception. Incipient cavitation would not affect the low to mid frequency results but rather provide increased levels from around 10 kHz upwards. It should be remembered that all of the pumps referred to used water as the pumping fluid.

FAN NOISE PREDICTION

Fan noise measurement

9.   The accuracy of fan sound level measurement is less than is generally believed. Under rigorous laboratory conditions the standard deviations of in-duct measurements are of the order of 2.5 dB. Noise measurements taken on site, where conditions may be far from ideal, are inherently less accurate An analysis of noise levels from a 3.66 m diameter cooling fan showed that the sound power level determined according to certain near field test procedures was not within 5 dB of the level determined by the more accurate far field hemispherical test. Thus the same unit can be rated differently according to the test procedure adopted.

10.   Noise measurements should be performed with the fans operating under a standard acoustic load impedance and this may require tests on each of the four combinations of ducted inlet/ducted outlet; free inlet/free outlet; free inlet/ducted outlet; ducted inlet/free outlet, if such configurations are possible. Providing a standard impedance requires strict attention to the test ducting and the use of transitions, and to the performance of terminator in preventing reflections Some form of airflow turbulence suppressor must be fitted to the microphone used for in-duct measurements.

11.   The above factors may imply a need to reappraise data currently used in design project work and necessitate the development of scaling laws enabling model testing to replace on-site measurement.

Simple prediction techniques - sound power levels

12.   Numerous empirical formulae exist for calculating fan sound power levels which require only a knowledge of overall parameters such as volume flowrate and fan pressure rise. Typical of such formulae are

$$L_W = 27.5 + 8 \log_{10} Q + 24 \log_{10} P \qquad (3)$$

and $L_W = 91 + 10 \log_{10} (QP) + 30 \log_{10} \frac{M}{M_{20}}. \qquad (4)$

The first of these formulae applies specifically to backward curved centrifugal fans (Ref. 1), while the second (Ref. 2) is claimed to be more general.

13.   Extensive measurements on a wide range of axial, mixed and radial-flow fans have yielded results showing that the ratio of the inlet or outlet sound power to the aerodynamic power is related to the fan tip speed by the expression

$$SWR = \frac{L_W}{QP} = 4 \times 10^{-10} (V_t)^2. \qquad (5)$$

Provided that the fan design conforms to the following criteria this relationship can be used to determine the inlet or outlet sound power to within ±3 dB. The essential restrictions are that

a   the fan tip speed is within the range of about 40-150 m/s. As the fan speed increases the exponent increases,

b   the fan is operating at its best efficiency duty point,

c   there is no excessive turbulence in the inlet airflow,

d   there is reasonable clearance between rotor and stator if fitted or between the rotor trailing edge and the cut-off if the fan is a centrifugal.

Once the overall sound power has been estimated a frequency spectrum can be predicted.

Estimation of frequency spectrum

14. If accurate noise measurements exist for one fan the spectrum of a geometrically similar unit may be derived using scaling factors.

15. The effects of size and speed have been evaluated (Ref. 1) for a range of geometrically similar centrifugal fans and the sound power level L_W was found to be proportional to $N^{5.6} D^{7.2}$. Using the fan affinity laws the following relationships may be derived to relate changes in sound power level to fan size and speed

$$\Delta L_W = L_{W,2} - L_{W,1} = 56 \log_{10} \frac{N_2}{N_1} + 72 \log_{10} \frac{D_2}{D_1}$$

$$= 28 \log_{10} \frac{P_2}{P_1} + 16 \log_{10} \frac{D_2}{D_1} \quad (6)$$

$$= 24 \log_{10} \frac{P_2}{P_1} + 8 \log_{10} \frac{Q_2}{Q_1}.$$

If no spectral data exist, estimates can be made using empirical correlations. For centrifugal fans it can be assumed that the frequency spectrum is linear, peaking at low frequency and reducing in amplitude at a rate of 3 dB per octave. Tonal effects are generally unimportant in centrifugal fans.

16. Axial-flow and mixed-flow fans are acoustically more complex than centrifugal fans with a large number of parameters influencing the noise generation and radiation, particularly of the tonal components.

17. Various methods are available for estimating the broad band noise component. One, developed at NEL, uses the correlation of data shown in Fig. 4 which relates to a wide variety of fans ranging from 0.6 m diameter up to 3.66 m diameter and includes both mixed and axial-flow fans, some with rotor only and others with a stator row. Frequency bands which contained dominant harmonic tones were excluded from the correlation, thus the correlation represents only the broad band components of the fan noise spectrum. The correlation refers to a fan operating at its peak aerodynamic efficiency duty and in 'aerodynamically good' conditions.

18. The frequency is non-dimensionalized with respect to the impeller tip chord and speed of sound, and the amplitude is in the form of a spectral density relative to the overall sound power level.

19. The overall inlet or outlet sound power level can be calculated using equation (5). Since this includes a tonal contribution, 2 dB should be subtracted implying that about half the energy is tonal. The broad band spectrum can be derived from Fig. 4, and if the frequency band containing the fundamental tone is raised to a level equal to (overall sound power level - 4 dB) a reasonably accurate estimate of the spectrum is obtained.

20. Wright (Ref. 3) presents an alternative way of predicting the broad band spectrum. The spectrum has a peak sound power level given by the empirical equation

$$L_W = -52.6 + 2|\alpha - 4| + 60 \log V_t + \left(20 - \frac{V_t}{4}\right) +$$

$$+20 \log D + 10 \log Bc + 10 \log \Delta f \; dB^* \quad (7)$$

at a frequency,

$$f_{peak} = V_{tip}/\text{tip chord} \quad (8)$$

and the spectrum shape is such that if the sound power is plotted in constant percentage bandwidths (for example $\frac{1}{3}$-octave) the levels are 30 dB below the peak value at $f_{peak}/8$ and 10 dB lower at $8 \times f_{peak}$. For low-speed fans the prediction is within 6 dB of the measured levels at frequencies above the peak frequency.

21. The prediction formulae are for fans operating at best efficiency flowrate. At other duties the overall sound power level generally rises (see Fig. 5), although complex changes occur in the frequency spectrum. In axial and mixed-flow fans there is a tendency for the blade passage tone radiated from the inlet to reduce as the flowrate is reduced while the tone level increases on the discharge side if stator vanes are fitted. Beyond the stall point there is a marked increase in the low frequency sound power levels.

NOISE CONTROL

Fan noise

22. Three broad aspects of the control of fan noise may be considered. These are

a   the design of the system and selection of the correct size and type of fan,

b   ensuring the fan has a good acoustic design, and,

c   the application of silencing techniques.

System considerations

23. Fan noise is, to a first approximation, proportional to fan air power. Therefore an object of system design should be to minimize the power requirements, for example by using larger diameter ducting to reduce flow velocities. In axial and mixed-flow fans the ineraction of the rotor and the inflow turbulence is generally the dominant source of the blade passage frequency tones.

---

*The term $(20 - V_t/4)$ is included only if it is positive

Particular attention should therefore be given to ensure that airflow into a fan is uniform and as free from turbulence as possible. Struts, supports, flow controls etc, all of which can generate wakes and turbulent flows should be downstream of a fan. Thus, for example, outlet guide vanes are prefereable to inlet guide vanes. If air is being drawn into a fan inlet or inlet duct the area around the inlet should be unobstructed with no objects closer than 1 diameter otherwise the flow could be asymmetric. If the intake is open to the atmosphere the effects of crosswinds on the intake should be minimized.

Selection of fan

24. Once the fan system has been designed the head rise and flowrate requirements can be specified and the appropriate type and size of fan chosen. The type of machine can be judged by calculating the non-dimensional specific speed, N_s.

In broad terms the specific speed will characterize the type of fan as follows.

$$N_s < 2 \quad \text{for centrifugal fans}$$

$$1.5 < N_s < 3.5 \text{ for mixed-flow fans}$$

$$2.5 < N_s \quad \text{for axial-flow fans.}$$

25. If the fan is axisymmetric the diameter, D, may be estimated using the expression

$$D = 2\left(\frac{gHQ}{\omega^3 K_L}\right)^{0.2} \qquad (9)$$

derived by Farrant (Ref. 4).

Values of the loading coefficient, K_L, typical of commerically available fans are in the range $K_L = 0.04$ to 0.08 though values up to almost 0.2 can be attained (Ref. 4). Using different values of ω, a range of fan diameters can be derived. Then, since the diameter, speed and air power are known, equation (5) will yield sound power levels for each choice of speed. If a chord length can be estimated, Fig. 4 can be used to calculate a frequency spectrum. For centrifugal fans an approximate tip speed can be calculated from

$$V_{tip} = \left(\frac{gH}{0.8 - 0.23N_s}\right)^{\frac{1}{2}}. \qquad (10)$$

It will then be clear how the noise levels will be affected by the choice of fan.

26. Many fans operate over a range of duties and a check must be made to ensure that the fan can satisfy all duty conditions. At low flowrates axial fans stall and the sound power can increase by 15-20 dB. At low flowrates centrifugal fans can develop rotating stall (Ref. 5) which may cause severe pulsations at around $\frac{2}{3}$ of shaft frequency.

Aspects of fan design

27. The best way of reducing fan noise is to design for the minimum tip speed which means increasing the aerodynamic loading on the blades. Acoustically this introduces many complex changes. With reduced velocities the magnitude of pressure fluctuations and the radiation efficiency is reduced. Since highly loaded blades

are less sensitive to inlet turbulence this source of noise is reduced. However both blade boundary layer noise and blade wakes are related to the drag coefficient which increases with aerodynamic loading and there is an increase in the relative strength of these components. To a limited extent drag can be lowered by choice of blade section while the effect of wake velocity deficit can be reduced by an increased rotor/stator spacing. Measured rates of change of tone level with axial spacings have ranged to as high as 24 dB per doubling of spacing but typical values are 4 dB per doubling of distance up to 1 chord spacing and 2 dB thereafter. Despite this widely known fact machines are still designed with minimal rotor-stator clearnace. One cooling fan unit on an electric motor being used at NEL had a clearance equivalent to 0.2 chord length. The sound pressure level 1.6 m from the unit was 101 dB re 20 $\mu N/m^2$. After removing the stators the noise level dropped to 88 dB re 20 $\mu N/m^2$.

28. Increasing the separation will not result in a change of the fundamental tone if inlet turbulence is the dominant noise source. This is probably why modal cut-off does not always appear to be successful. If the rotor wake/stator blade interaction is the dominant tonal source, then there are combinations of rotor and stator blade numbers which should generate non-propagating tones.

29. Experimentally modal cut-off has been demonstrated on several fans at NEL and numbers are now chosen to ensure cut-off even though the benefit may be masked by turbulence. Fig. 6 presents some experimental results demonstrating the effectiveness of cut-off, the results being in line with theoretical predictions. One feature which is most strongly brought out is that the practice whereby rotor and stator numbers are often chosen so that they differ by only one or two will invariably lead to an efficiently propagating blade tone.

Fan noise silencing

30. Silencers which depend on reactive elements are generally too bulky for fan systems and most fan silencers consist of a porous lining placed in the flow path which removes sound energy directly by internal friction within the porous layer. On occasions the flow may be dusty, contaminated with chemicals, laden with steam or very hot, thereby effectively eliminating most conventional silencers. In such circumstances the use of barriers can provide a simple and effective solution. In one case the erection of deflectors on top of four fan discharge stacks reduced far field noise by some 5-8 dBA, sufficient to overcome a neighbourhood noise problem.

Noise control within pumps

31. It is thought that pumps react to inlet flow distortions in a manner similar to fans although it is by no means certain that pumps are primarily dipole sources. Current scale effect investigations at NEL will identify the size and speed scale law and indicate whether pumps may be of monopole or dipole type. However even considering the pump as a monopole and considering the noise source to be regions of separated flow, the size of these regions would vary with the inlet flow incidence variations in a manner similar to that producing fluctuating lift.

32. Tests were carried out at NEL on a two-stage

pump which was fitted with a set of inlet guide vanes of a given solidity set close to the first stage inlet. The distance between the vane trailing edge and the impeller leading edge represented about 5 per cent of the guide vane chord length. Since the guide vanes were considered necessary for smooth and efficient pump operation at low flows their hydrodynamic effect was retained by preserving the original solidity while replacing the long blades by many short blades whose leading edge was in the same position as before. Thus the stator/rotor clearance was significantly increased. The change in outlet noise level at the design flow is shown in Fig. 7 where a 10 dB broad band noise reduction can be clearly seen up to 800 Hz. This reduction extends to higher frequencies. Similar reductions in outlet noise level were shown for both higher and lower flowrates. At the inlet a smaller and less consistent reduction was evident at all flows. Reductions in inlet turbulence were clearly beneficial.

33. Tonal noise is often a problem in pump systems and to date no systematic information is available for noise reduction methods. Two methods, sometimes advocated in fan systems, which might be applicable to pumps are increased volute to impeller clearance and the use of asymmetric impellers.

34. In the same pump as described above, the cutwater clearance was increased from about 8 per cent to about 14 per cent in an attempt to reduce noise at blade passing frequency. The pump pressure rise reduced by around 1.5 per cent whilst the outlet tonal noise rose by 5 dB at design flow and reduced by similar amounts at off-design. The broad band noise remained largely similar and at the inlet the spectrum was unchanged.

35. Sometimes in fan systems where a pronounced tone exists, use of non-symmetrically spaced rotor blades depresses the blade passing frequency at the expense of introducing sidebands above and below the blade rotational harmonic. In one attempt to reduce tonal noise an asymmetric pump impeller was designed and manufactured to the same duty as the original symmetric unit and a comparison of noise output made. Fig. 8 shows a sample of a spectrogram measured at the outlet for design flow where the typical result is apparent. Rather than depress the main blade passing frequency level, tones have been introduced at every harmonic from the 120-780 Hz shown with no advantage even in broad band terms.

36. The two attempts at noise reduction lead to the suggestion that the principal method of noise control must be through reduction of impeller tip speed and good hydrodynamic design of impeller inlet considering noise generation.

CONCLUSIONS

37. Data on fluid-borne noise from pumps and fans have been correlated to provide a simple means of estimating overall sound power and sound pressure levels. Correlations of the frequency spectra enable estimates to be made of the spectral distribution without detailed knowledge of the geometrical design.

In both fans and pumps tip speed reductions will effect noise reductions. Duct systems should be designed for minimum pressure losses and fans and pumps chosen so that they operate at best efficiency flowrate.

Turbulence in the inlet should be minimized as it is a source of tonal and broad band noise.

On fans increasing the impeller outlet clearance reduced noise, but tests on a pump showed an increase of noise with increased cutwater clearance.

Use of asymmetric blading on a pump increased rather than reduced the tonal components.

Modal cut-off is effective in fans if inlet turbulence is suppressed.

ACKNOWLEDGEMENTS

The authors wish to thank the United Kingdom Atomic Energy Authority for permission to present certain data, and all colleagues at NEL who assisted in the experimental work.

This paper is contributed with the permission of the Director of the National Engineering Laboratory, Department of Industry. It is Crown copyright.

REFERENCES

1. DEEPROSE W.M. and BROOKS J.M. Effect of scale on fan noise generation of backward curved centrifugal fans. NEL Report No 512. East Kilbride, Glasgow: National Engineering Laboratory, 1972.

2. ERSKINE J.B. and BRUNT J. Prediction and control of noise in fan installations. Conference on Vibrations and Noise in Fan, Pump and Compressor Installations. Instn mech. Engrs, September 1975.

3. WRIGHT S.E. The acoustic spectrum of axial flow machines. Journal of Sound and Vibration, 1976, 45(2), 165-223.

4. FARRANT P.F. A method for selecting casings for mixed flow pumps and fans. International Conference on Design and Operation of Pumps and Turbines. National Engineering Laboratory, East Kilbride, September 1976.

5. BOLTON A.N. Pressure pulsations and rotating stall in centrifugal fans. Conference on Vibrations and Noise in Fan, Pump and Compressor Installations. Instn mech. Engrs, September 1975.

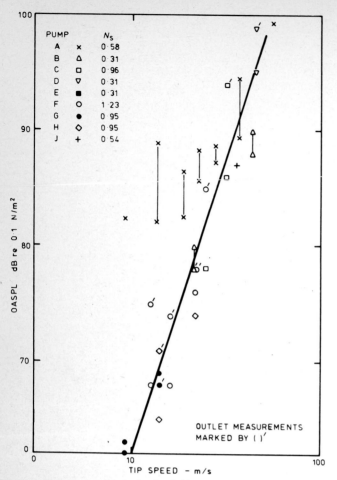

Fig. 1: Overall sound pressure level of a range of pumps as a function of tip speed

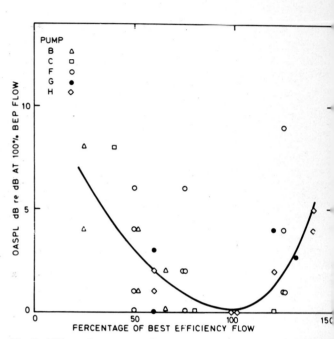

Fig. 2: Effect of percentage flow on overall sound pressure level for a number of pumps

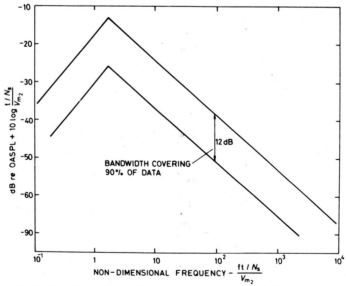

Fig. 3: Non-dimensional frequency spectrum for centrifugal pumps

48

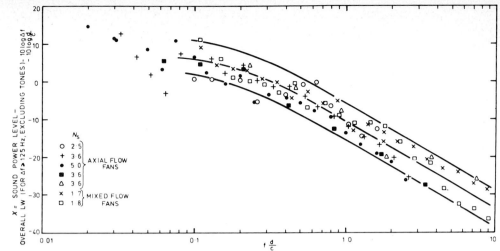

Fig. 4: Correlation of axial and mixed-flow fan broad band noise

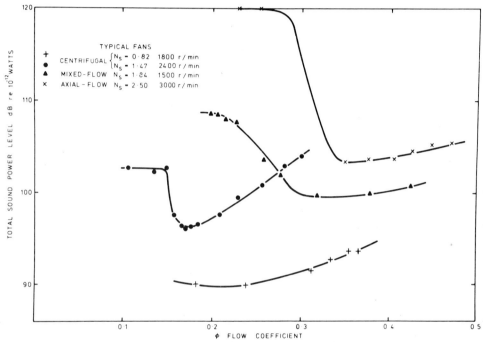

Fig. 5: Effect of flowrate on overall sound power level for various fans

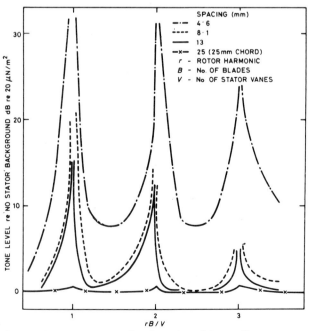

Fig. 6: Experimental verification of modal cut-off

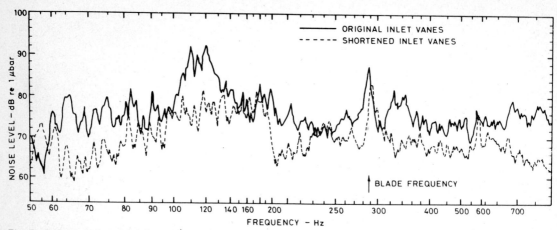

Fig. 7: Effect of changing inlet guide vanes on pump discharge noise

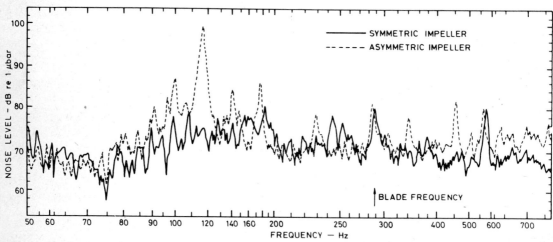

Fig. 8: Effect of asymmetric impeller on pump discharge noise

C252/77

NOISE CONTROL STRATEGY FOR MANUFACTURERS AND USERS

M.F. RUSSELL, MSc
Research Department, CAV Limited, London

The MS of this paper was received at the Institution on 18 July 1977 and accepted for publication on 8 August 1977

SYNOPSIS   As a manufacturer of fans and pumps, CAV Limited has to control the noise from its products to satisfy its customers' requirements; and a product noise control strategy has been evolved. These requirements are based, partly, on legislation controlling the maximum noise levels emitted by road vehicles. As a noise-conscious purchaser of fans, pumps and compressors, for use in its factories, the company has had to develop a strategy to control the noise levels of new plant and machinery, and to embark on a programme to mitigate the noise from existing equipment. The technical problems associated with both activities are dealt with by a central specialist noise control group, which applies the techniques and equipment developed to control noise from products, to the plant and equipment which the company purchases.

INTRODUCTION

1.   CAV Limited is concerned with noise in its factories arising from production equipment, including pumps, fans and compressors; it is concerned also with noise generated by its own products, including pumps and fans, which are affected by increasingly stringent legislation on the noise of the road vehicles to which they are fitted. The pumps inject fuel into diesel engines in pulses lasting for between one half and three milliseconds, at pressures of up to 15,000 lb/sq.in. This requirement places very sudden and severe loads upon the mechanism, which may generate noise as clearances are taken up, or as the structure responds to the sudden and transient forces. The fans are part of alternators and motors fitted to a variety of vehicles; such fans can be a major noise sources of the alternators at high rotational speeds. Techniques developed to analyse and control noise generated by the company's products, have been a valuable part of the noise control programme for production equipment.

STRATEGY FOR CONTROLLING NOISE FROM PRODUCTS

Fuel injection pumps

2.   There are two principal reasons for endeavouring to reduce the noise from fuel injection pumps:-

(1)   To improve the "quality" of the noise emitted by diesel powered vehicles at and near idling conditions.

(2)   To maintain the present situation, where fuel injection pumps are not contributing significantly to the noise from diesel engines at rated load and speed, when quieter engines are developed to meet the increasingly stringent legislation for road vehicles.

3.   In some cases noise control treatments have been devised at the design stage of new products, a few years in advance of legislation. When products have been established for a number of years, it has been possible to apply noise control treatments when the design is developed to meet new requirements, particularly where changes in tolerances or production specifications give a benefit in noise without incurring large additional expense. Considerable economies have been made by incorporating noise control measures when production machinery is changed or renewed.

4.   This implies an anticipative research and development programme of some years duration to diagnose the sources of vibration or noise within the products, and the noise radiating surfaces, followed by modelling of the process to investigate modifications to the noise-generating process. It also implies a continuity of target, and of policy, to achieve this economically.

Fans

5.   The reduction of noise from alternator fans became more important when it became clear that the market was demanding more output from each frame size. This could be achieved by altering the windings at the expense of the speed at which the alternator started to generate (cutting-in speed). This necessitated a alternator to engine speed ratio so the alternator would run at such a speed that it would generate whilst the engine was idling. Noise from fans rises rapidly with speed, and it was felt that noise could become a barrier to sales if it started to contribute significantly to the noise from a vehicle engine, whilst the vehicle was tested in a drive-by test such as ISO R362, (ref. 1), or similar tests, (ref. 2), required by legislation. At the same time it was felt desirable to remove the whine from the fan to improve the "quality" or "acceptability" of the noise emitted.

Legislation

6.   The drive-by test specified in ISO R362

specifies a test site incorporating a 20m long track (Fig. 1) which is approached in the second gear of a 4-speed box at ¾ governed speed, or ¾ maximum power speed or 50 km/hr, whichever is the lowest. Alternatively, for vehicles with many gear ratios, a gear is chosen such that when the engine is rotating at ¾ of its maximum power speed, the vehicle is travelling at or near 50 km/hr. The vehicle is accelerated maximally over the test track, and then allowed to decelerate with idling fuelling. The noise level is measured by a microphone placed opposite the mid point of the track, 7.5m from the centre line of the vehicle. The implications of legislation based on such a test have been examined for a variety of vehicle classes, and one typical study is shown in Figure 2.

7. The existing and proposed legislation for vehicle noise in the E.E.C. is shown as the top, stepped, line in Figure 2. There is no firm proposal for a further reduction around 1985, but the dotted line represents an estimate of likely future reductions in noise. This highlights one of the problems in trying to meet noise legislation, in that the forward planning necessary to do it economically seems to exceed the forward planning of the legislators by a considerable margin. For the purpose of establishing noise targets it has been assumed that the engine of the truck might contribute half of the overall noise intensity at the measurement microphone. The maximum level for fuel injection equipment and alternators is 7 to 10 decibels below the noise from the engine, where the contribution to the overall noise from the engine becomes insignificant. The figures relate to a truck greater than 3500 kilogrammes in weight fitted with an engine with less than 200 bhp installed power, which is intended to operate within the E.E.C. countries.

8. Noise levels of the products which might be fitted to such a truck are shown in the right hand column, based upon measurements of sound power in an absorbent-lined room. The levels quoted refer to the noise emitted when the engine is running at its maximum rated load and speed, with the alternator generating full load, which is somewhat pessimistic. Forecasts have been made of the possible noise problems for all the company's products affected by such legislation. This exercise was first carried out in 1971 and has been updated at intervals to take account of changes in legislation. As a result, the product noise control programme has kept well ahead of the requirements brought about by legislation, so far.

Noise control of products at the design stage

9. There exists a time period during the evolution of a design when it is possible to influence the design not only to avoid obvious faults like large clearances in mechanisms but also to introduce positive measures to restrict the noise which might otherwise be generated from the device. Very often the time period for such advice is quite short and there may not be enough time to do a comprehensive study of the mechanism of noise generation. When such situations can be foreseen then there is a clear case for a research study to provide the bulk of the requisite data in advance of the requirement.

10. One example of the preparation of data for the design of fans is shown in Figures 3 and 4. Figure 3 shows how the noise level of a particular machine increased with speed with no electrical load: the fan noise increases with speed at a rate which fits very well with the equation

$$SPL \text{ (in decibels)} = X + 53 \log_{10} \left(\frac{N}{1000}\right)$$

where N is the rotational speed in revs/min

and X is the noise level at 1000 revs/min
1 metre from the machine

11. Figure 4 shows the individual and combined effects of fan and end shield on the value of X and upon the airflow. The original design was a shrouded centrifugal fan with short blades, randomly spaced about its periphery, and this was replaced by a fan with an increased blade area with blade lengths adjusted to maintain good airflow into the fan from the alternator rotor. The curve in the left hand side of Figure 4 shows the trade-off between noise and airflow for the new fan at various fan diameters, since the noise decreases as the gap between the fan blade tips and the bearing support spider increases. Comparing the standard fan in the original end shield (S in Fig. 4), with the large diameter of the new fan (Δ), there is an increase in the airflow with the new fan due to a larger blade area, coupled with a small reduction in noise. As the new fan is reduced in diameter the airflow falls off as the noise decreases to a point where at the same airflow as the standard fan there is a 6 dB(A) reduction in noise. Thus by improving the fan in the original prototype machine it is possible to obtain either a small reduction in noise with some increase in airflow or a 6 dB(A) reduction in noise for the same airflow.

12. It seemed likely that further reductions in noise might be made if the endshield design could be altered as well: and accordingly a new end shield was constructed which contained four directional outlets, to see what the benefits would be available if one of the design constraint (the ability to operate equally well in either direction of rotation), could be relaxed. The original fan gave more airflow in the end shield with directional ports (S^1 in Fig. 4). The directional end shield increased the airflow from the new fan, with increased blade area, considerably. A smaller fan of the same design could give the same airflow as the original end shield with the largest diameter fan. At 16.5 cubic feet per minute the directional, ported end shield with smaller fan gave a 5 dB(A) reduction compared with the largest diameter of the improved fan in the standard end shield, or if compared with the original design a reduction of 7.5 dB(A) with an increase in airflow. With the smallest fan which was tried the noise was only 1 dB above the level with no fan, which does not leave a great deal of room for improvement. When larger diameter fans were used in the directional-ported end shield the airflow was improved considerably with a correspondingly increasing penalty in noise. Armed with the knowledge of these trade-offs between noise and airflow the designer can alter his specification for different applications. The improvements to fan and end shield took 5 months to develop and drew on four previous investigations of fan noise.

Introduction of noise control measures into a well established product

13. Once a product is in production, further constraints in noise control techniques arise. The changes must be suitable for using with existing manufacturing plant, or the designer must wait until a plant change is contemplated.

14. An example of this type arose with the company's DPA rotary fuel injection pump.

15. Noise from the DPA pump does not rise as rapidly with speed as combustion and mechanical noise sources in diesel engines (Ref. 3 and 4). At the high speeds called for in the ISO R362 drive-by test, this pump makes no significant contribution to the engine noise. At and near idling, the pump can be heard on some engines although it is at least 5 dB(A) below the engine noise. It is desirable to improve the "quality" of the noise at idling by removing as many as possible of the impacts which occur within the pump.

16. There are at least seven possible impacts in the shaft and pumping mechanism. With some of these impacts, it is difficult to tell how the vibration is transmitted to the noise radiating surfaces of the pump. Not all these impacts occur in all pumps; and their relative severity is altered by the dynamics of the drive. They have been ranked in order of importance for typical pumps by relating the changes in relative timing of the measured impacts, to the changing sequence of events inside the pump as the pump speed and fuelling are changed. Fig. 5 shows the sequence of events and the impacts measured with an accelerometer on the pump housing.

17. The modes of vibration of the pump housing and governor cover have been determined by exciting the pump with a sinusoidal force. These modes were subsequently confirmed by pulsed laser holography as the pump was vibrating at each of its natural frequencies in turn (ref. 5). One result of this work is shown in Fig. 6, where the 'panel movement' of one of the thinner parts of the housing can be seen from the elliptical fringes. After the diagnosis had established the major sources of structure-borne noise and the normal modes which radiated most noise, a definition of the transmission "paths" of the vibration inside the pump was attempted. This part of the diagnosis proved more difficult since vibration can be transmitted from the impacting members via the bearings, the moving parts, the cam restraint and the fuel with which the pump is filled.

18. Three approaches have been tried:-

(a) Minimise or eliminate all impacts at source.

(b) Build isolation into structure to prevent vibration of impacting parts flowing into pump housing.

(c) Stiffen pump housing to inhibit certain noise-radiating modes of vibration.

19. Each of these approaches has been shown to give reductions of approximately 5 dB(A). Attempts to obtain further reductions by a single approach have led to impractical modifications, which adversely affect pump performance or present serious manufacturing problems. For example there are further modifications to the pump housing which would considerably reduce its response to vibration, but these are inadmissible as they would incur huge re-tooling costs.

20. A composite package is being developed in which each stage of development provides a quieter pump. The stages introduced so far are:-

Stage 1   Stiffer housing with reduced response which will still go down the transfer line.

Stage 2   Stage 1 + adoption of single-piece drive shaft with a close-fitting splined coupling.

21. The reduction in noise achieved by Stage 2 can be seen in Fig. 7.

STRATEGY FOR CONTROLLING NOISE IN COMPANY FACTORIES

Purchasing standards

22. When purchasing new plant and equipment it is necessary to specify the maximum permissible noise levels to ensure that the risk of hearing damage to those who will work at or near the machinery is minimised. This involves a decision on the target level for the factory noise. It is widely accepted that noise above 90 dB(A) can be damaging after long periods of continuous exposure, but there is still much controversy over the level at which damage starts to occur in the working population. Taking 90 dB(A) as an interim target, new machines would have to be less than 85 dB(A), and in the case of small machines such as single spindle automatic lathes, 84 dB(A), when measured individually in their final location. (ref. 6)

23. The Company standard for new plant and machinery specifies that the noise level should be less than 83 dB(A) when measured in the suppliers premises, in its noisiest normal operating condition. This makes a small allowance for the possible inaccuracies in sound level meters and the difference in acoustics between the suppliers premises and the final location. Clearly it is highly desirable that the machinery should be measured in precisely-defined acoustic conditions, but in practice the size of the Company's factory, the size, shape and disposition of machines within it, and the acoustic absorption properties of the structural materials used for the buildings, are remarkably similar to the suppliers assembly area or despatch bay.

24. Furthermore the Company Standard is written to measure noise in the near field, where it affects the people who are close to it, and where the influence of room acoustics is small in most cases.

25. The Company has a suite of Noise Standards (ref. 7) covering noise surveys of premises, maximum levels for noise in Company premises, action to be taken when that level is exceeded, noise levels for new plant and machinery and methods of measurement of new plant and machinery. The purchasing procedures include measurement of new plant and equipment at the suppliers premises before the machine is accepted. These Standards are designed to identify noise problems and their cause in a systematic manner so that solutions can be devised, either by the local Works or Production

Engineers who have been trained in noise measurement and control, or by the specialist noise control centre.

26. The maximum permissible noise level for new machines helps to control the noise from pumps, fans and compressors used in workshops but there are additional problems of ventilation fans and oil coolers mounted on factory roofs or outside the factory, and also ventilation equipment used in or near offices. Such problems demand individual consideration, and standards such as the machine noise standard are not very useful. Specifying noise levels at the factory perimeter has its pitfalls too and may involve needless expense. The approach under consideration at present, is to allow a limited degree of individual adjustment in order to meet standards like ISO R1996 (Ref. 8) and BS 4142 (Ref. 9). This involves specifying the criteria to be met at the residences most likely to be affected by the noise source under examination and working back to a noise requirement for the individual piece of equipment. In order to make the standard workably simple, the effects of shielding, focussing, ground effects, etc. have to be left out of this initial assessment. These effects can be considered when the specialist noise centre investigates those fans or compressors which are shown to be likely to give excessive noise as a result of the initial assessment. At this stage a more accurate assessment may be made to provide targets for noise control measures which may need to be fitted. It remains to be proven how well this approach will work in practice, and how much training of Works and Production Engineers will be required to make the initial assessments realistic.

27. All the above procedures are aimed at incorporating noise criteria sufficiently early in the specification-design-development-production sequence to obviate the need for remedial noise control. In practice this state of affairs is hard to achieve and hard to maintain and problems arise with existing products and plant which require a combination of a thorough understanding of the noise generating mechanism and an imaginative approach to noise control.

Remedial noise control during operation

28. An example of the remedial action which is sometimes necessary, occurred in a workshop where cars were repaired and tuned. The workshop noise level was approximately 65 dB(A). Some weeks after a new battery fume extraction fan had been installed, a whine developed which grew in volume until it disturbed the engine tuners who were using the noise from the engines as a measure of regularity. The whine was traced to the impeller shaft bearings, which appeared to function adequately even when noisy. The bearings were replaced, (not an easy job as the fan is mounted 15 ft. up, on a wall), but several weeks later the whine returned, and the specialist noise control centre were asked to help.

29. The loading on the ball bearings and their life expectancy with this loading was estimated to be ten weeks which was close to the actual period of use before the whine developed. Considerable modification to the bearing plinth would have been required to fit larger bearings to the existing plinth, so a frame was constructed to carry larger bearings which overhung the plinth, to avoid altering the relative height of the impeller shaft and the plinth. The frame was isolated from the plinth with cork-filled rubber pads designed to prevent the bearing-frequency forces from exciting the resonant plinth while providing adequate stiffness to hold the shaft in position. The design of the frame and the noise reduction is shown in Fig. 8. The calculated life of the new bearings was two years. Three years after the frame was installed, the belt-end bearing was found to be worn but no increase in noise was observed and no complaints had been received from the engine tuners.

DEVELOPING AN EFFECTIVE NOISE CONTROL STRATEGY

30. To achieve cost effective noise control at source it is essential to have:

(1) A detailed understanding of the noise-generating mechanism.

(2) A detailed understanding of the functioning of the noise-generating device and the factors which affect its performance.

(3) A clearly defined and justified noise level target.

31. Experience has confirmed, many times, the need for a complete diagnosis and ranking of structure-borne and air-borne noise sources together with the modes of vibration of panels and covers which radiate structure-borne noise. These are the first steps to understanding the noise generating process. The understanding of the functioning of a machine is that which a designer or development engineer acquires after working on the machine during its evolution. This is particularly difficult for the user of a machine to acquire, and this renders him less able to evaluate the practical problems which might result from implementation of some noise control treatments. Armed with this knowledge it is often possible to devise several noise control schemes, each giving different final noise levels; some which may be used in combination to build up a "noise control package". The selection of the package depends upon what can be manufactured economically, and the target noise level which has to be achieved. Since different target levels often involve large differences in cost, it is important to forecast the requirements sufficiently far in advance to develop the noise packages in time.

32. A successful programme requires forecasting future noise level requirements at least five years ahead, to enable the necessary research work and development of noise control packages to be completed ahead of the requirement. In the event of introduction having to await the next production tooling change, the total time may exceed five years.

33. Investment in equipment and expertise for product noise control has produced a bonus in terms of ability to quieten supplier's products, using techniques and equipment necessary for the CAV product programme.

ACKNOWLEDGEMENT

34. The author thanks the directors of Lucas
Industries and CAV Limited for permission to
publish this paper, and he gratefully acknowledges
the support and assistance of Mr D Amos,
Mr A Herbert, and Miss C Shaw.

REFERENCES

1. International Organisation for Standardisation,
"Measurement of Noise Emitted by Motor Vehicles"
Recommendation R362.

2. British Standards Institution, "Method for the
Measurement of Noise from Motor Vehicles" BS 3425;
1966.

3. RUSSELL M.F., and HERBERT A.J., "Identification
and Modelling of Rotary Fuel Injection Pump
Noise Processes", Society of Automotive Engineers,
Diesel Engine Noise Conference, Milwaukee,
Sept. 1975 - Paper No. 750803.

4. RUSSELL M.F., "Reduction of Noise Emissions
from Diesel Engine Surfaces", Society of Automotive
Engineers Congress, Detroit - Jan. 1972 - Paper
No. 720135.

5. RUSSELL M.F., and HERBERT A.J., "Holography",
CAV Research Department Note N176, Jan. 1976.

6. RUSSELL M.F., and MAY S.P., "Machinery Noise:
the Users Viewpoint", Proc. of 17th International
Machine Tool Design and Research Conference,
University of Birmingham, Sept. 1976. MacMillan
Press (Revised version in Machinery and Production
Engineering, 2nd Feb. 1977, pp 98-102).

7. Lucas Group Standards Dept., "Lucas Standards
for Noise Control and Hearing Conservation", Lucas
Industries, Great King Street, Birmingham.

8. International Standards Organisation, "Assessment of Noise with Respect to Community Response",
ISO/R1996 - 1971(E).

9. British Standards Institution, "Method for
Rating Industrial Noise Affecting Mixed Residential
and Industrial Areas", BS4142, 1967, Amended 1975.

APPROACH at 50km/hr | ACCELERATION at FULL FUELLING to attain SPEED s | DECELERATION

Fig. 1: Vehicle drive-by test, ISO R362

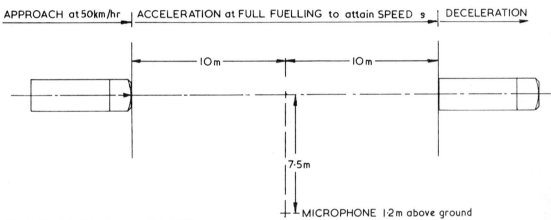

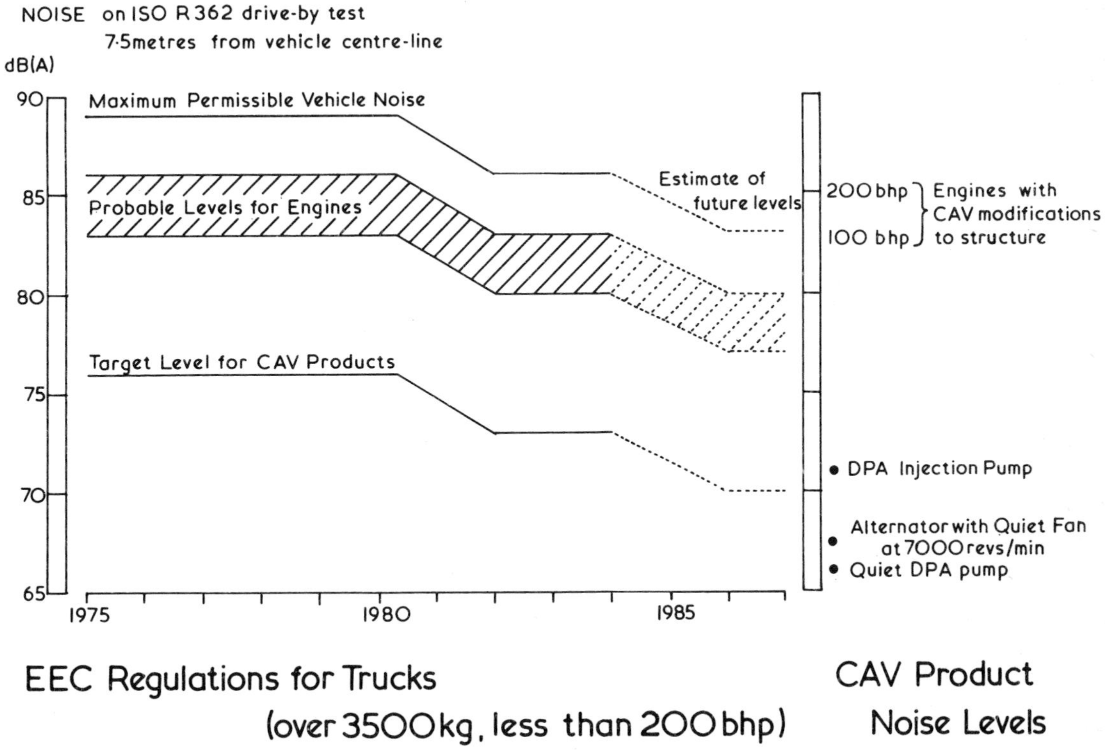

Fig. 2: Comparison of EEC truck regulations and CAV product noise levels

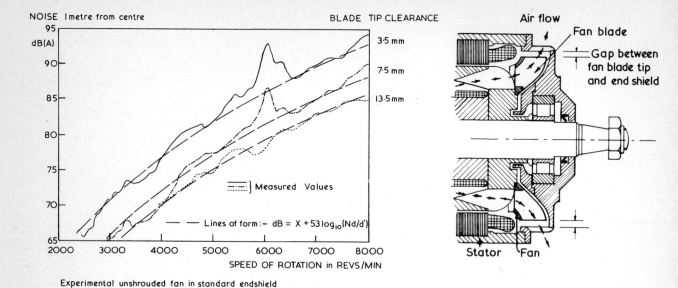

Fig. 3: Noise from fans in an electrical machine

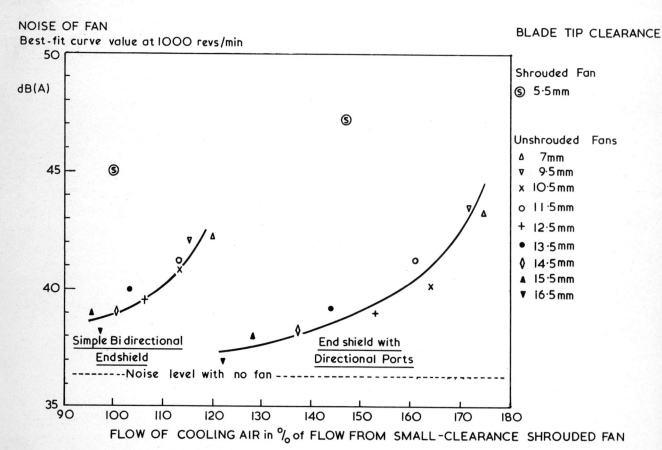

Fig. 4: Trade-off between noise and air flow for an electrical machine fan

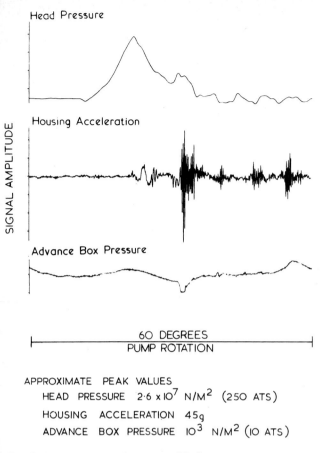

SIGNAL AMPLITUDE

Head Pressure

Housing Acceleration

Advance Box Pressure

60 DEGREES
PUMP ROTATION

APPROXIMATE PEAK VALUES
HEAD PRESSURE $2 \cdot 6 \times 10^7$ N/M^2 (250 ATS)
HOUSING ACCELERATION 45g
ADVANCE BOX PRESSURE 10^3 N/M^2 (10 ATS)

Fig. 5: Data recordings from a modified pump

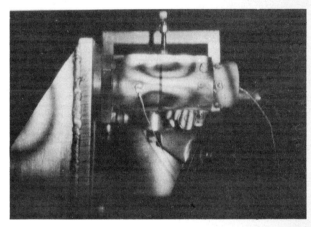

Fig. 6: Modes of vibration of a pump housing and governor cover

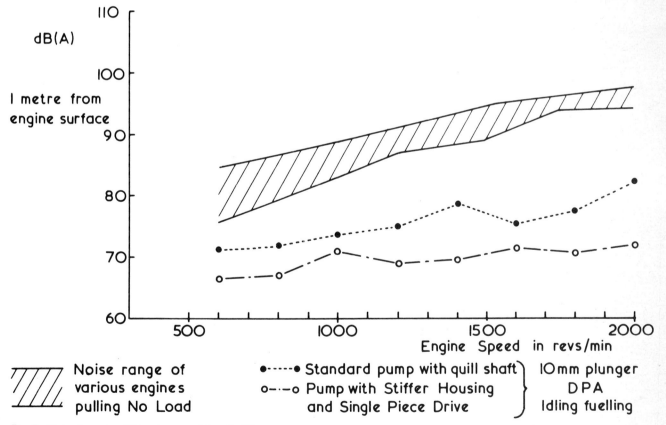

Fig. 7: Noise from a quietened pump at low fuelling

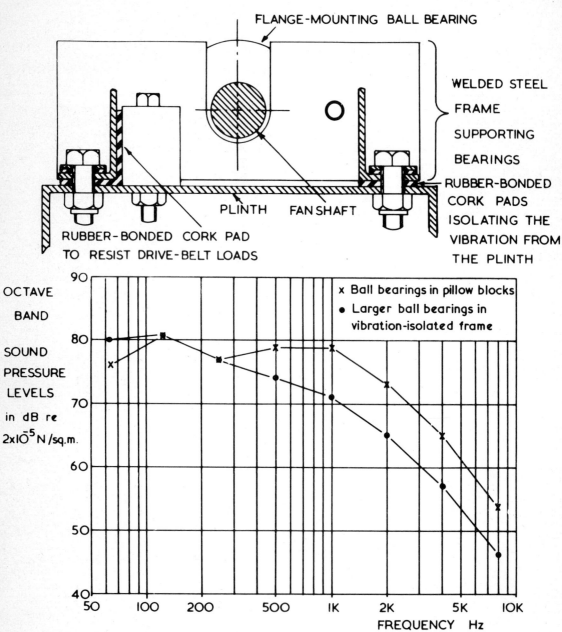

Fig. 8: Treatment for persistently noisy ball bearings

58

C254/77

FAN NOISE RESEARCH AND ITS IMPLICATION FOR NOISE LEVEL PREDICTIONS

B.D. MUGRIDGE, BTech, MSc, PhD
Formerly Institute of Sound and Vibration Research, University of Southampton

The MS for this paper was received at the Institution on 6th May 1977 and accepted for publication on 11th August 1977

SYNOPSIS   Fan noise research is briefly reviewed and the fundamental sound generating mechanisms are described.   Both broadband and rotational components of the radiated noise spectrum are considered and some guidelines given for noise reduction at source.   The fundamental concepts are used to derive general formulae for predicting the broadband noise of industrial axial and centrifugal flow fans.

1. INTRODUCTION

1. Most of the research into sound generated aerodynamically, particularly by rotating blades, has been linked with developments in the aircraft industry. The initial interest was in the sound radiated by propeller blades but in more recent times the main effort has been towards understanding both the fan and jet noise of turbomachinery. Axial flow fans have therefore been studied in great depth and the basic noise generating mechanisms are now well established. The information is being used to estimate, and where possible, control the noise of fans used in other important areas such as heavy industry and the heating and ventilating field. This paper briefly describes the early research efforts that have led to this improved knowledge although space does not permit inclusion of all the important contributions that have made possible our appreciation of the noise generating processes. The paper then continues by considering the techniques that have been developed for predicting the total sound power of relatively low speed industrial fans. In principle this is an easier task than would be encountered in the aircraft engine field since the mechanical simplicity and lower shaft speeds of the industrial designs eliminate some of the more complex noise generating mechanisms.

2. NOTATION

A Fan inlet area
C_d Blade drag coefficient
c Blade chord
D Fan diameter
f Frequency (Hz)
$G_v(\omega)$ Velocity spectral density
l_s Spanwise correlation length
N Shaft speed (RPM)
n Harmonic number
P_s Fan static pressure head (mm wg)
Q Air flow rate (m^3/min)
U_T Fan tip speed
U_r Mean velocity relative to blade
W Sound power
η_s Efficiency (fractional)
ρ Air density
ϕ Flow coefficient $(= Q/AU_T)$

ψ Pressure coefficient $(P_s/\rho U_T^2)$
Ω Shaft speed (rads/sec)
ω Frequency (rads/sec)

3. BASIC NOISE MECHANISMS

2. In 1918 Lanchester recognised that the major source of propeller noise was the steady rotation of the blade force field. This source was subsequently examined by a number of investigators and Gutin (1) determined the amplitude of the radiated sound as a function of the propeller steady loading. Lighthill's (2) general theory of the sound generated by turbulent flow was then extended by Curle (3) who demonstrated theoretically that the interaction of an unsteady flow field with a rigid surface produced surface pressure fluctuations that could be considered as the origin of a more efficient noise source. A fan or helicopter blade rotating in free space (or in a duct) was thus shown to radiate noise from both the steady and fluctuating lift forces (4). Ffowcs Williams et al (5) then confirmed the intuitive idea that the steady loading on the rotor blades and any periodically induced unsteady loading would generate a discrete frequency noise spectrum and that any randomly induced force field due to, say, the interaction of the blade with turbulent flow, would produce a broadband noise spectrum. These were significant contributions towards explaining the observed noise spectrum characteristics of an axial flow machine, propeller or fan.

3. Predicting the amplitude of the total radiated sound has always depended on an accurate assessment of the unsteady flow to blade interaction. Although Curle's hypothesis of an acoustic dipole source related to surface force fluctuations is the most easily understood source model, it is also possible to consider the direct radiation of the turbulent flow field whose acoustic properties are radically altered by the presence of the immersed solid surface. Turbulence is a relatively inefficient radiator of sound since the phasing between individual components of the flow field tends to be self cancelling. The effect of inserting a solid surface into this flow is to produce 'scattering' which reduces this cancellation and therefore

causes an increase in the radiation efficiency (6).

4. The scattered turbulence idea is useful for calculating the noise radiated from the blade trailing edge but it is often more convenient to use the acoustic dipole model since the surface pressure distribution may be more readily measured and the sound radiation computed. Under some conditions the pressure field may even be calculated. If the fan blade chord is less than the acoustic wavelength then the surface pressure distribution caused by the interaction of the blade with a convected non-uniform flow may be calculated from unsteady airfoil theory. This theory has been recently extended by Kemp et al (7) and may be used to determine the fluctuating force on a fan blade for both periodic and random flow disturbances. Inserting this force and a spanwise correlation length into Curle's basic equation gives the source strength from which the radiated sound levels may be calculated.

3.1 Rotational noise

5. The steady thrust force on a propeller blade radiates sound because of its acceleration relative to the observer. On a multi-bladed rotor disc this is an inefficient radiation process since the force pattern rotates at subsonic speed. For equally spaced rotor blades the sound occurs at the blade passing frequency and its harmonics. When the rotor operates in a disturbed velocity field typical of, say, the wake flow from a stator row then the rotating blades experience periodic forces. If the stationary blade row consists of V equally spaced identical vanes then, as shown by Tyler et al (8), the rotor force field at the blade passing frequency contains components which rotate at $B/(B-sV)$ times the rotor speed, where s is any integer positive or negative. Thus if sV is close to the blade number B part of the force field can rotate supersonically even when the rotor speed is well below sonic. It is well known (Ref. 9) that there is a rapid increase in the radiated sound power when the acoustic mode, $m=B-sV$, begins to rotate supersonically and it is for this reason that periodic force fluctuations are the major contributors to fan discrete tone noise.

6. The distortion mode sV may be considered as the λ th harmonic of any spatial inlet flow non-uniformity which generates the λ th loading harmonic on any one rotor blade. At the rotational frequency $n\Omega$ there may be a number of loading harmonics, λ, that contribute to the radiated sound waves, in fact the number of contributing harmonics is increased as the fan speed Ω increases. Under some conditions one loading harmonic may contribute to more than one rotational harmonic. For equally spaced rotor blades all the sound waves cancel except at multiples of the blade passing frequency. The tones are often heard as a sharp whistling sound which can be very annoying.

7. If the dominant interaction is that of a stationary blade row with a rotating flow distortion produced typically by the upstream rotor wake flow, then each loading harmonic is given by a wake harmonic and corresponds to a sound harmonic. The total radiation from the stator row at a sound harmonic (e.g. $\lambda = nB$) then depends on the contributions from all the stator blades. This

total sound may be reduced by judicious unequal circumferential spacing of adjacent stator blades (Ref. 10).

8. The techniques for reducing noise from rotor-stator interactions are well established. The axial spacings between rows must be at least one chord in order to reduce the strength of the interaction, and the blade numbers must be chosen to give a high mode number, m, to take full advantage of its reduced radiation efficiency. It is more difficult to reduce the tone noise caused by interactions of the rotor with an inlet flow distortion field since one does not have a free choice of the loading harmonics λ. One method is to unequally space the rotor blades since this will alter the preference for radiation at the blade passing frequency at the expense of additional radiation at other multiples of the shaft rotational frequency, see Figure 1.

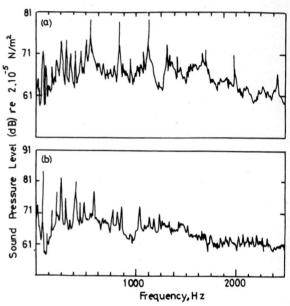

Fig. 1: Noise spectra of axial fan in highly distorted inlet flow field

(a) symmetrical fan
(b) fan with non-symmetrical blade spacing

This method is sometimes very effective particularly at low shaft speeds but if the inlet flow field is very distorted then the multiplicity of loading harmonics λ give a spectrum rich in rotational harmonics where the dB(A) rating is often no better than the original equally spaced rotor (Ref. 11).

3.2 Broadband noise

9. The most common sources of broadband noise in fans are interactions between the blades and oncoming turbulent flow, the effects of the blade turbulent boundary layers and the influence of the blade to duct tip clearance. Little is known of the latter although the author did determine that the noise radiation from this source was influenced by the three dimensional flows induced by the blade loading, tip clearance and duct boundary layer thickness (Ref. 12). The inlet turbulence noise may be calculated using unsteady lift theory for the fluctuating force strength. The main difference between this turbulent model and the periodic interactions mentioned in

Section 3.1 is the spectral distribution of the upstream flow field and the magnitude of the spanwise correlation length of the induced forces.

10.   Turbulent boundary layer noise is less easily predicted. One technique is to use Curle's analysis and measured boundary layer induced pressures. Alternatively the scattered turbulence model previously discussed can be used if the strength of the turbulence in the vicinity of the blade trailing edge is known or can be estimated. Unfortunately a recent comparison (Ref. 13) showed that these two models did not give the same value of radiated sound intensity since the force and turbulent fields were not known with sufficient accuracy. To add to the present difficulties Archibald (14) has published some interesting data on a possible mechanism whereby the sound field is controlled by a boundary layer flow instability which is driven by an acoustic feedback loop. This is evidently one area where further research is needed.

4.   FAN NOISE PREDICTIONS

4.1   Axial flow fans

11.   It is not difficult to apply the general principles of aerodynamic sound generation and derive equations for the total fan sound power level as a function of the various parameters defining the steady and unsteady flows passing through the rotor disc. For example, the inlet flow interactions give the proportionality,

$$W(\omega) \propto U_r^4 \; D^2 \; G_v(\omega) \cdot \left(\frac{\omega \cdot l_s}{U_r}\right)$$

where $G_v(\omega)$ is the spectral density (proportional to velocity squared) of the inlet turbulent flow or a statement of the harmonic content of the periodic disturbances. The final bracket includes the influence of the spanwise correlation length of the inlet flow, a blade load weighting factor deduced from unsteady aerofoil theory that accounts for a finite chordwise correlation length, and the time differential of the blade force field. (See, for example, Ref. 11). The relative velocity U_r is determined from the blade velocity triangles.

12.   Using Curle's force model the fan residual broadband noise produced by the blade turbulent boundary layers and which in some extreme flow conditions may dominate the total noise spectrum, can be expressed in the following form,

$$W(\omega) \propto U_r^6 \; D^2 (c_f)^q$$

Here c_f is the local skin friction coefficient beneath the boundary layer flow. Normally this friction coefficient is expressed in terms of the overall drag coefficient C_d of the rotor blade row and is therefore related to the fan aerodynamic efficiency. The power index q is usually determined experimentally. The broad peak of the radiated spectrum appears to be controlled by a Strouhal number, $St = (f \Theta)/U_r$ based on the momentum thickness of the turbulent wake flow, Θ. Since this thickness and the blade drag coefficient C_d are directly related, then a fan with a low aerodynamic efficiency (high drag coefficient C_d) will radiate predominantly low frequency noise whilst a fan with a much higher efficiency would have the main acoustic energy in the mid to high

frequency bands.

13.   The analysis may be extended to express the sound power more explicitly in terms of the fan pumping capabilities and the nondimensional coefficients ϕ and ψ that define the off design performance of a particular fan design. Mellin (15) has already shown that the residual broadband noise of a single axial rotor can be expressed in terms of the flow coefficient ϕ. More recently Longhouse (16) has shown that varying the flow coefficient of an isolated rotor by flow control changed the relative contributions of the tone and broadband noise components to the overall spectrum. The tonal components, caused mainly by flow distortions, dominated the noise spectrum at high flow coefficients (low disc loading) whilst the broadband noise became dominant at low flow coefficients (high disc loading). In particular the exponent of noise versus rotor tip speed was mechanism dependent. This means that as the fan operating point moves along the flow coefficient-pressure coefficient curve of a particular blade design the sound power law varied from the U_r^6 to some other value which could be as low as $U_r^{4.5}$.

14.   Using this approach Mugridge (17) attempted to derive theoretical expressions for the overall broadband sound power of axial flow industrial fans based on the fundamental models previously described but written in terms of the basic fan parameters P_s, Q, ϕ, ψ, and static efficiency η_s. These parameters encompass speed and size effects and may also be combined to give power consumption. The results gave the following two expressions of proportionality between the spectral sound power $W(\omega)$ and the fan parameters.

For inlet turbulence interactions

$$W(\omega) \propto Q \; P_s^{2.5} \; \left(\frac{\phi}{\psi^{2.5}}\right) \cdot \left(1 + \phi^2 - \psi + \frac{\psi^2}{2}\right)^2 \cdot G_v(\omega) \quad (1)$$

For turbulent boundary layer noise

$$W(\omega) \propto Q \; P_s^{2.5} \; \left(\frac{1}{\psi}\right)^{1.5} \left(\frac{1 - \eta_s}{\eta_s}\right) \cdot \left(1 + \phi^2 - \psi + \frac{\psi^2}{2}\right)^{1.5} \quad (2)$$

15.   The relative importance of these two mechanisms to the total fan noise is obviously dependent on the flow environment, but these expressions were correlated with sound power data for fans operated under normal manufacturers test conditions. Within the accuracy of the experimental data (approx. ± 2dB), which did not distinguish between the two different mechanisms, the octave band sound power level (PWL) radiated from either the inlet or outlet of a ducted axial flow fan with minimum tip clearance (e.g. 0.5%) could be adequately described by the expression

$$PWL(f) = 23 + 7\log_{10}Q + 25\log_{10}P_s + 10\log_{10}Z +$$
$$F_2 \cdot \text{dB re } 10^{-12}W \quad (3)$$

where $Z = \left(\frac{1 - \eta_s}{\eta_s}\right) \cdot \left(\frac{1}{\psi}\right)^{1.5} \cdot \left(1 + \phi^2 - \psi + \frac{\psi^2}{2}\right)^{1.5}$

The octave band spectral factor F_2 is shown in Figure 2.

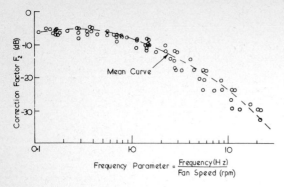

Fig. 2: Axial fan octave sound power spectrum shape F_2

It should be noted that for any one fan design (or family of similar designs) the coefficients ϕ and ψ are uniquely related. If the fan design remains constant then a single $\phi - \psi$ curve describes its performance. Increasing the fan speed N or diameter D at <u>constant efficiency</u> η_s then corresponds to a single point on the $\phi - \psi$ curve. Therefore at constant efficiency Equations (1) and (2) may be written as $W(\omega) \propto N^6 D^8$. However, the equations, and experimental data, show that the N^6 law need not hold if the fan operating conditions involve moving <u>along</u> the $\phi - \psi$ curve.

16. Where possible the author prefers to examine all fan noise data in terms of the separate mechanisms since different fan installations can have a marked effect on the radiated sound. For example, the suitability of Equation (3), based on Equation (2), to predict the sound power of axial fans tested in standard ducted calibration test rigs having downstream flow throttling, implies that under these conditions turbulent boundary layer effects are of major importance. However, Equation (3) would underestimate the measured data if the fan is installed in a highly turbulent or distorted flow field. Nevertheless, it is interesting that if Equation (3) is written in terms of the tip speed and power consumption then it gives a very similar result to the empirically derived overall noise prediction formula developed by Hawes (18) for fans operating in highly turbulent inlet flows. Actually if Equation (1) and (2) are used it is fairly straightforward to show that for a typical axial flow fan design there is a similar variation in normalised sound power as a function of flow coefficient, $\phi \gg \phi_{opt}$, for both inlet turbulence and boundary layer interaction noise. Thus although the frequency spectra would generally be different for the two mechanisms the <u>overall</u> noise levels could show the same parametric variation.

17. The general conclusions are that for a given pumping requirement Q, P_s, the broadband noise is reduced if the static efficiency and/or the pressure coefficient is increased (tip speed decreased). Alternatively for high flow applications (cooling fans) the pressure head (system pressure losses) should be kept to a minimum.

18. The prediction of blade passing frequency noise for rotor-stator interactions can be carried out with reasonable accuracy in the manner described in Reference (10). As mentioned earlier, noise reduction is achieved by a suitable selection of blade numbers, adequate axial spacing and,

if possible, non radial stator vanes. Interactions between a rotor and an inlet flow non-uniformity should be analysed for each particular situation. In most cases the blade passing tones may be predicted with fair accuracy using line dipole theory (Ref. 19) as shown in Figure 3.

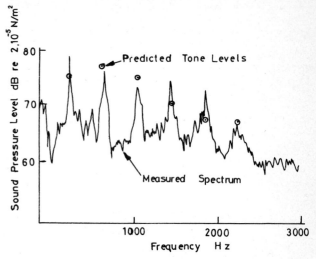

Fig. 3: Blade passing frequency noise levels for axial fan with inlet flow distortion

19. An automotive cooling fan is a good example where accurate noise prediction depends on a clear understanding of the flow environment and of the basic equations describing the noise source. In this instance it is more convenient to revert to first principles and rewrite the broadband and discrete tone noise equations as a function of <u>measured</u> inlet flow distortions and turbulence correlation lengths. If this is done for each particular type of environment then quite good noise predictions can be achieved. Reference (11) describes such an exercise and Figure (4) indicates the degree of accuracy obtained.

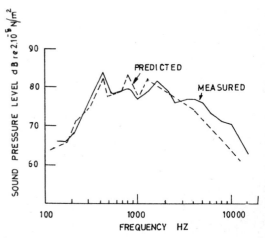

Fig. 4: One third octave band noise spectrum of axial flow cooling fan

4.2 Centrifugal fan design

20. Very little work has been done on the aerodynamic noise of centrifugal fans and blowers even though they are used extensively in heating

and ventilating plant. Because they operate at
low Mach numbers with impeller configurations
that are susceptible to partial stall, most of
the sound power is generated in the lower frequency bands and therefore influenced by the
acoustic reflection properties of the inlet and
outlet duct and the impeller housing. The discrete frequency tone at the blade passing frequency
is primarily caused by aerodynamic interaction
between the impeller wake flow and the volute cut
off. Neise (20) has collated data that demonstrates the importance of the cut off spacing and
radius, and cut off or impeller inclination, on
the magnitude of the blade passing tone. In a
typical industrial fan design this tone is about
5dB higher than the broadband noise level within
the appropriate octave band (usually 500 Hz).

21. It is not always clear whether the broadband
noise is created predominantly within the impeller
or the volute since the velocity distribution
varies considerably with different fan designs.
The forward curve bladed centrifugal impeller
acts mainly as a momentum gaining device and
relies on the volute for efficient conversion of
the high resulting kinetic energy into useful
pressure energy. In backward curved bladed fans
most of the useful work occurs in the impeller
itself with the volute acting basically as a
collector. Fortunately it is somewhat easier to
derive a broadband sound power law for the centrifugal fan than the axial type since the flow
pattern within the centrifugal design is quite
often determined by the housing and impeller geometry and is not too sensitive to upstream flow
turbulence. The exception is for highly throttled
fans where the damper is located immediately
adjacent to the fan housing. Except for the last
condition the broadband noise output may be real-
istically predicted by the method given in
Reference (17). The octave band PWL is given by
the Equation,

$$PWL(f) = 33 + 7\log_{10}Q + 25\log_{10}P_s + 10\log_{10}Z +$$

$$F_1 \cdot \text{dB re } 10^{-12} \text{ W}$$

The spectral correction factor F_1 is zero at
125 Hz and falls off at $3\frac{1}{2}$ dB for every subsequent
doubling of frequency (i.e. $F_1 = -7$dB at 500Hz and
$F_1 = -14$dB at 2000Hz). This applies to both backward and forward bladed impellers since the influence of the different designs is catered for
through the different pumping capabilities and ϕ,
ψ characteristics. The suitability of the term Z
to express the variation of the normalised sound
power as a function of the flow and pressure coefficient, is indicated in Figure 5 which compares
Z with a typical measured variation for a forward
vaned centrifugal impeller. Similar comparisons
can be obtained for backward vaned impellers. The
formula indicates that minimum noise occurs at
maximum efficiency ($\phi \rightarrow \phi_{opt}$) and with the minimum
pressure head required to deliver the air flow
rate.

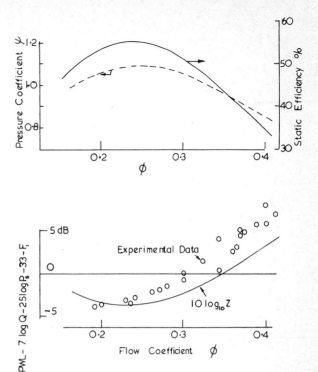

Fig. 5: Forward vaned centrifugal impeller. Variation of fan
efficiency, pressure coefficient, and normalized sound
power with flow coefficient

5. CONCLUSIONS

22. A brief description has been given of the
progress achieved in fan noise research. Not all
of the important contributions have been included
but the interested reader is referred to the exc-
ellent review paper by Morfey (21). Some of the
noise reduction methods described have been known
for some time and there are instances where the
techniques have been applied with considerable
success. Unfortunately this has not occurred in
all sectors of the fan industry partly due to the
increased cost of manufacture that some of the
modifications involve and partly because not all
of the laboratory measured reductions are realised
in practical situations. It is expected that this
situation will improve as a consequence of the
general tightening up on environmental pollution
in and around industry. Noise reduction may even
come as an indirect result of improvements in industrial fan design. The recent introduction of
variable pitch axial fan blades to improve the off
design aerodynamic performance should also improve
the noise characteristics.

APPENDIX

REFERENCES

1. GUTIN,L. 'On the sound field of a rotating airscrew', Zhurnal lekhnicheskoi fiziki 1936 6, pp. 899-909. Translated as 1948 NACA TM 1195.

2. LIGHTHILL,M.J. 'On sound generated aerodynamically. I. General theory, Proc.R.Soc. 1952 A 211, pp. 564-587.

3. CURLE,N. 'The influence of solid boundaries on aerodynamic sound, Proc.R.Soc. 1955 A 231, pp. 505-514.

4. LOWSON,M.V. 'The sound field for singularities in motion, Proc.R.Soc. 1965 A 286, pp. 559-572.

5. FFOWCS WILLIAMS,J.E. and HAWKINGS,D.L. 'Theory relating to the noise of rotating machinery', J.Sound & Vib. 1960 10, pp. 10-21.

6. FFOWCS WILLIAMS,J.E. and HALL,L.H. 'Aerodynamic sound generation by turbulent flow in the vicinity of a scattering half plane', J.Fluid Mech. 1970 40, pp. 657-670.

7. KEMP,N.H. and HOMICZ,G. 'Approximate unsteady thin airfoil theory for subsonic flow', AIAA J. 1976 14, pp. 1083-1089.

8. TYLER,J.M. and SOFRIN,T.G. 'Axial flow compressor noise studies', Soc.Automot.Engr. Trans. 1962 70, pp. 309-332.

9. MORFEY,C.L. 'A note on the radiation efficiency of acoustic duct modes', J.Sound & Vib. 1969 9, pp. 367-372.

10. DUNCAN,P.E. and DAWSON,B. 'Reduction of interaction tones from axial flow fans by non-uniform distributions of the stator vanes', J.Sound & Vib. 1975 38, pp. 357-371.

11. MUGRIDGE,B.D. 'The noise of cooling fans used in heavy automotive vehicles', J.Sound & Vib. 1976 44, pp. 349-367.

12. MUGRIDGE,B.D. and MORFEY,C.L. 'Sources of noise in axial flow fans', J.Acoust.Soc.Amer. 1972 51, pp. 1411-1426.

13. MUGRIDGE,B.D. 'Sound generated by flow over surfaces with particular reference to industrial fan noise', 1975 Machinery Noise and Environment Colloquium No. 2, Paris.

14. ARCHIBALD,F.S. 'The laminar boundary layer instability excitation of an acoustic resonance', J.Sound & Vib. 1975 38, pp. 387-402.

15. MELLIN,R.C. 'Selection of minimum noise fans for a given pumping requirement', Noise Control Eng. 1975 4, pp. 35-45.

16. LONGHOUSE,R.E. 'Noise mechanism separation and design considerations for low tip-speed, axial flow fans', J.Sound & Vib. 1976 48, pp. 461-474.

17. MUGRIDGE,B.D. 'Noise characteristics of axial and centrifugal fans as used in industry', Shock & Vib.Dig. 1975 7, pp. 93-107.

18. HAWES,S.P. 'Modern fan technology', Automot. Eng. 1976 pp. 15-17.

19. MUGRIDGE,B.D. 'Axial flow fan noise caused by inlet flow distortion', J.Sound & Vib. 1975 40, pp. 497-512.

20. NEISE,W. 'Noise reduction in centrifugal fans-a literature survey', J.Sound & Vib. 1976 45.

21. MORFEY,C.L. 'Rotating blades and aerodynamic sound', J.Sound & Vib. 1973 28, pp. 587-617.

C255/77

A PUMP MANUFACTURERS APPROACH TO NOISE EMISSION LIMITATIONS

J.K. MACDONALD, CEng, MIMechE and B. REIL, BSc(Eng)
Mather & Platt Ltd., Manchester

The MS of this paper was received at the Institution on 5 July 1977 and accepted for publication on 15 August 1977

SYNOPSIS   An increasing awareness of the harmful effects of noisy environments is forcing the equipment manufacturer to comply with predetermined noise emission limitations.   The centrifugal pump presents difficulties in obtaining satisfactory data for contractual acceptance and design of future products.   The design and manufacture of equipment to comply with customer specifications and statutory noise legislation are discussed, together with the measuring techniques employed and the difficulties of acoustic isolation of ancilliary equipment.   Where extra noise reduction is required, external noise suppression may be used as an economic solution and the design and effectiveness of suitable equipment is described.

INTRODUCTION

1.   Ten years ago, few pump users' purchase specifications required measurement of the generated sound levels of our pumps and virtually none specified maximum acceptable levels of noise emission.

2.   During the intervening years, the situation has generally changed and now predicted or guaranteed noise data is requested, along with the more usual specifications such as speed, generated head and flow.

3.   The code of practice for employee protection (1) and more recently, the Health and Safety Act (2) have forced plant operators and employers to take an active interest in noise control.

4.   In existing noisy working areas the sound level must be reduced where possible by silencing machinery and by applying some acoustic treatment to the building. If the noise cannot be reduced sufficiently by these methods personal ear protection may have to be provided.

5.   However, for new installations accurate noise emission data must be available for all significant noise from new machines if an attempt is to be made to maintain an acceptable noise level throughout the plant.   A pump manufacturer must be able to provide as much noise data about his products as possible and ensure that this information is accurate and relevant to the operational environment of the machinery.

NOISE MEASUREMENT IN MANUFACTURER'S PLANT

6.   It is the current policy in the Authors' Company and many other pump manufacturers to measure the noise emission from pumps as they undergo their hydraulic performance tests.   With the smaller lower powered units, measurements of noise can be made accurately, but with the high pressure, high flow pumps the necessary ancillary equipment often produces a higher noise level than the pump itself.

7.   In many cases the main single source of noise is the discharge valve where the water power from the pump is dissipated.   If a conventional valve is used in this application, the turbulence and cavitation generated downstream of the valve induces significant noise and vibrations throughout the pipework system which travels back to the pump casing and forward along the discharge pipework.   It is extremely difficult to dissociate this valve noise from the pump noise.

8.   To reduce the noise emitted from the test system discharge valve, the Authors' Company now use their own design of valve which does not create the violent turbulence which is normally associated with conventional pressure reducing valves.   Such an approach is imperative when testing high speed boiler feed pumps generating heads in excess of 2100 m where noise levels associated with conventional test loops can not only inhibit normal working in the manufacturers' plant but may lead to complaints by local residents.   The latter can be a serious problem if it is necessary to test such machines during the night to comply with power restrictions imposed by the electricity supply authority.   Ideally, such a valve must be mounted as far from the pump as possible leading to considerable capital investment in high pressure pipework.   Even under these circumstances it will be necessary to make allowance for the valve noise when assessing the noise characteristic of the pump under test.

9.   A further significant source of noise can be the drive unit, usually an electric motor, which may be run in isolation to confirm that its noise emission is as specified by its manufacturer.   Booster pumps, when required, may be mounted close to the pump under test but these are relatively quiet and do not usually present a problem.   A gearbox or variable speed coupling used between the motor and pump results in additional complications which must be taken into account.   Finally, the larger pump units require considerable quantities of lubricating oil and the auxiliary oil pumps required for this purpose can also be noisy.

10.   The noise from induction motors alters very little with load and their noise emission can be measured by running the motor uncoupled.   However,

some variable speed machines with rotary converters can present isolation problems. A similar condition exists with a gearbox or variable speed coupling, where the noise may vary with speed and load. In some cases it is only feasible to supply the pump user with noise measurements of the complete pumping set, the results effectively determining the position of any outstanding noise generator but not accurately rating the individual sources.

11.   Whilst fulfilling the customers requirements for the particular unit to enable him to assess his needs for acoustic isolation, this information is of limited use to the pump manufacturers in predicting the noise levels of a similar pump with an alternative drive system. For this it is necessary to isolate the pump from extraneous noise sources.

12.   The optimum method of achieving this isolation is to position the pump in an anechoic chamber with drive and ancillary equipment positioned remotely from the chamber. However, such an approach is not feasible on a production test facility for large or high power pumps and an alternative method is necessary for this class of machine.

13.   Screens may be erected to reduce the drive system and discharge valve noise. The use of 19mm chipboard or plywood with a 50mm thick layer of absorbent material on the pump side is a simple expedient which has been sucessfully employed, allowing accurate direct measurement of the noise emitted from the pump. Such an approach is not always feasible as a short drive shaft may make the fitting of such screens between pump and driver difficult.

14.   At present, most noise measurements are taken using the 'near field' method with the microphone 1m from the surface of the machine. It is necessary to ensure that there are no other large noise sources nearby and that there are no sound reflections from the building structure to affect the meter readings. Measurements of the octave band sound pressure level are taken at points around the pump and the average values calculated from these readings. If more detailed information is required a narrow band analysis is made, which may be used to locate the source of the noise and assist efforts to reduce it.

15.   Measurements of the velocity of the surface vibration have been taken to try to establish if a direct comparison could be made between this and the generated sound pressure level. Whereas some relationship was found with velocity measurements taken at a selected point on one pump, other points on the same pump and an identical point on another pump failed to give any similarity. We have assumed that this was caused by the manufacturing variations in the pump casing.

16.   The pump test area of the Authors' Company has two large pits which were constructed as flow measuring tanks but are now redundant. We have examined the possibilities of using these as acoustic test chambers where we could measure the noise radiated from the pump isolated from the noise generated by valve, motor and other ancillary equipment. When not in use the facility would not cause any obstruction to our Production Test Department.

17.   From calculations it was found that the size of an anechoic chamber which could be constructed in the larger pit would limit the test facility to our range of smaller production machines. By making the walls of the chamber moderately absorbent, the reflected noise could be reduced sufficiently to allow 'near field' noise measurements to be made in relative isolation from outside sources, but a more accurate alternative would be to make all the pit walls and ceiling highly reflective and try to achieve a diffused sound field.

18.   The dimensions of the large pit conformed roughly to the preferred ratios of side lengths for reverberation chambers, and its volume should give a reasonable frequency performance down to the 125Hz octave band. Noise with pronounced tonal qualities generally precludes the use of reverberation rooms but pump noise is usually broad band in nature making this system the most effective.

19.   The smaller of the two pits may be used as an isolation chamber to house the motor and a lay-shaft would be fitted through the separating wall. Both pump and motor may be mounted on vibration isolators and the suction and discharge pipework acoustically lagged and isolated where it passes through the chamber ceiling.

20.   Very large pumps and possibly ones with large motor gearbox drives will still be too big to be accommodated in this test facility but it would provide more accurate measuring facilities for a wide range of our production machines.

NOISE MEASUREMENT ON SITE

21.   A pump operating against a system pressure on site eliminates the noise problems associated with a pressure breakdown valve. However, when trying to isolate the noise, site locations usually present even greater noise measurement problems than those found on the manufacturer's test bed. Once in service the control of the pump will be governed by its duty requirements and it may not be possible to operate it at the required test load. Auxiliary systems and adjacent machines which may themselves be noise sources can prove as difficult to isolate as the discharge valve. In many cases site environments are unsuitable for noise tests since low ceilings, reflective walls and insufficient access to take measurements are often limiting factors, making site measurements difficult to interpret. It is the Authors' experience that limited information about the noise characteristics of pumps is obtained from site measurement. There are some exceptions to this general experience as detailed in Test Example 1.

Test Example 1

22.   Two single stage centrifugal pumps located at a water pumping station had been fitted with new impellers to change their hydraulic characterisitics and when the first modified unit was commissioned it produced an unacceptably high noise level.

23.   The pumps could be run individually without other noisy machinery nearby. The suction and discharge pipes disappeared below the floor level a short distance from the pump casing. There was

sufficient room around the machine to take noise readings and there were no significant noise reflections from the building structure. Measurements of the noise were taken with one pump working at its full duty load (Fig. 1.). The results of the octave band analysis suggested that the noise could be reduced by enlarging the clearance between the impeller and diffuser. This was increased from 1mm to $8\frac{1}{2}$mm by machining out the diffuser.

24. The readings of the pump sound pressure level were repeated with this increased clearance but under the same operating conditions. The octave analysis showed that the noise levels at the basic blade passing frequence (125Hz) were virtually unchanged but at higher frequencies where noise would be generated by the impeller/diffuser passing frequency (750Hz) the levels were reduce by about 5dB. The total 'A' weighted levels and noise rating were reduced by a similar value.

THE USE OF ACOUSTIC ENCLOSURES

25. The acoustic specifications of some pump users require such low levels of noise that conventional pumps will not comply with the limits set. In these cases the pump manufacturer is forced to offer some form of acoustic barrier to surround the noise source and attenuate the direct radiation. Fig. 3. illustrates the difference between actual noise from a pump unit and typical specification

26. The type of barrier used has to be derived from a study of the attenuation required, the nature of the noise sources and the physical size of the machine.

27. If a low level of attenuation is required surface lagging of the noisiest sections of the machine with a proprietary noise material may be all that is required. When one part of the machine is outstandingly noisy it may be possible to design a small enclosure around this source and so reduce the overall radiated levels. However, when the attenuation levels required are more than 10dB a total enclosure of pump, coupling and motor will probably be necessary.

28. Surface lagging of machinery gives poor results unless the whole of the radiating surface can be covered. The thermal insulating jackets of boiler feed pumps, which may be thought to be good acoustic insulation due to their construction (fibre blanket with steel outer cladding), have no measurable effect on the radiated noise as a large part of the pump structure is left uncovered. To be effective, surface lagging must give total cover and be carefully fitted. If it is disturbed when maintenance work is carried out on the machine and not correctly replaced its effect can be greatly reduced. It is most suitable for use on ducts and pipes where little or no access is required and where there is little risk of the total cover of the lagging being broken.

29. Maintenance access must be carefully considered when a small enclosure is to be designed for a part of the machine system. In this case the whole enclosure will be removed and refitted each time repair work is carried out.

30. The enclosure must be easily removed and replaced, as sections which are difficult to refit will soon be 'lost' or modified to ease the repair work and

the acoustic performance of the modified enclosure may then be well below its design specification.

31. Large enclosures also have many problems. Whilst internal access may be available for minor routine work, large sections of the structure must be removable for major repairs. Extra services will be needed inside the enclosure such as lighting and electrical power for hand tools. Fire protection may be required especially if the pumped fluid is inflammable. Doors and access panels must have good airtight seals and all service connections passing through the enclosure must be well sealed and vibration isolated. Instrument monitoring panels can be installed inside the enclosure and viewed through double glazed windows.

32. The walls and roof of a large enclosure made from 70-100mm thick composite panels and properly designed should give a noise level reduction of about 30dB. However, the ultimate effect will be influenced by installation details over which the machinery supplier often has no direct control. The noise attenuation of the enclosure can be greatly reduced if no attempt is made to stop the direct noise transmission through the machine mountings into the building structure. For the same reason, vibration isolation must be achieved on the suction and discharge pipes which should also be fitted with surface lagging to reduce direct noise radiation.

33. Large enclosures are probably of greatest value in reducing noise levels some distance away from the machine such as in nearby offices or even in neighbouring residential areas. The total overall noise exposure to machine supervisory staff will probably not be reduced.

34. A large enclosure for a pump set will double the floor area needed and will cost about 7% of the price of the set.

Test Example 2

35. Complaints were received from a customer about the noise from a boiler feed pump set. Noise surveys indicated that the main noise source was the epicyclic gearbox between the pump and motor. When running at full load and speed the gears emitted a high noise level in the 2kHz octave band.

36. Re-design of the gearbox was not considered to be an economic solution. The fitting of an overall enclosure was not practical as the induction motor and pump were relatively quiet. To reduce the noise a small acoustic enclosure was designed to fit around the gearobx, which was built and tested on a pump set at the manufacturing plant.

37. The enclosure was constructed from 2.5mm steel coated on both sides with a heavy layer of polythene which acted as a damping agent. The inner surfaces were lined with a 50mm thick sound absorbing plastic foam and the whole structure was isolated from the base and from the oil pipes with a rubberised cork material. Joints in the structure of the enclosure were sealed with R. T. V. silicone rubber which could be easily removed and renewed during maintenance.

38. During the Works test the enclosure gave a reduction of 6dB in the level of 'A' weighted noise.

Most of this was in the high frequency octave bands which made the noise less annoying. The peak in the octave band analysis (Fig. 2.) was reduced by the enclosure and whilst the overall noise level around the pump set was still high, a worthwhile reduction was produced in the particularly irritating narrow band noise from the gearbox. The attenuation achieved on site was further improved as the bottom of the gearbox could not be shielded during works testing. Site conditions allowed for this defect to be rectified.

CONCLUSION

39.   This paper outlines the problems experienced by the pump manufacturer to comply with the increased demand made by pump users for machinery which meets stringent noise limitations. Over recent years pump speeds have increased and weights have decreased to produce more cost effective designs, the penalty of such an approach being the generation of higher noise levels from both the pumps and drive equipment. Recent Government legislation is designed to ensure that suppliers will strive to reduce the noise levels generated by their equipment. This will be achieved either by the selection of heavier, lower speed equipment or the provision of acoustic enclosures for high speed equipment. The parameters which govern noise emission from pumps are generally understood, but it is the Authors' opinion that only marginal improvement may be achieved by redesign of the basic product. Such improvements are automatically achieved by following good hydraulic and mechanical design practice.

40.   The result of pursuing the objective of reducing noise levels of pumping plant and systems is to increase the pump users capital outlay in purchasing suitable equipment.

REFERENCES

1.   Code of Practice for the Protection of Employed Persons from noise.

2.   Health and Safety at Work etc., Act 1974.

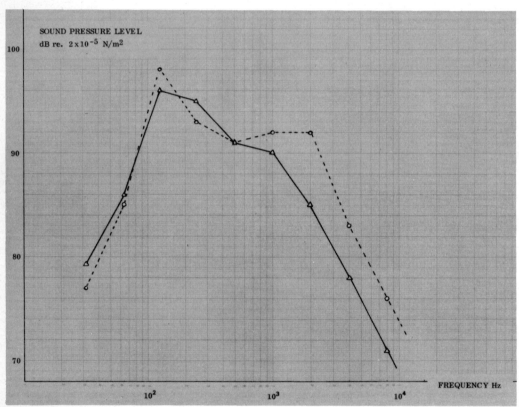

o 1mm radial clearance

▲ 8½mm radial clearance

Fig. 1: Reduction in pump noise level by increased clearance between impeller and diffuser

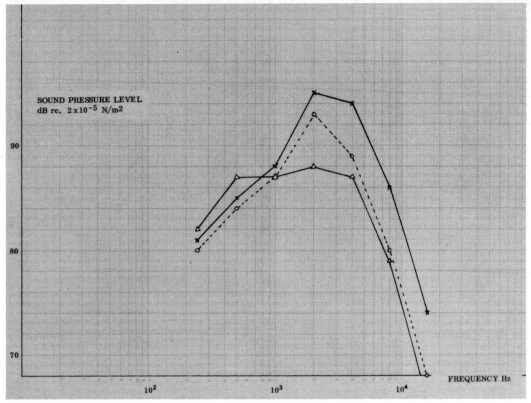

x Standard machine
o Plastic coated steel enclosure
△ as o but with 50mm absorbent lining added

Fig. 2: Reduction in noise level by local enclosure around source

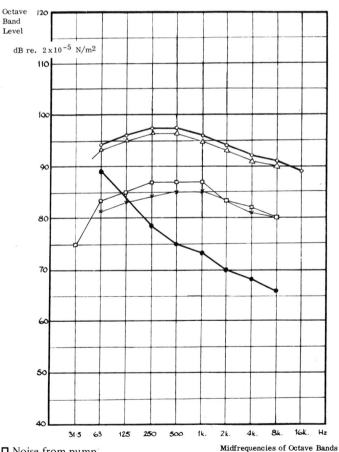

□ Noise from pump
x Noise from motor
△ Noise from gear box
o Maximum total level
● Specification limit

Fig. 3: Specified noise reductions where silencing is achieved by total enclosure